NEW Global ICT-based Business Models

by Peter Lindgren
(editor)

River Publishers

Aalborg

Published, sold and distributed by:
River Publishers
PO box 1657
Algade 42
9000 Aalborg
Denmark
Tel.: +4536953197

ISBN: 978-87-92329-76-9
© 2011 River Publishers

By members of the NEWGIBM research group:

Aalborg University, Center for Industrial Production
Fibigerstræde 16, DK-9220 Aalborg SV, Denmark
- Anders Drejer, Professor
- Jacob Høj Jørgensen, Doctoral Research Student
- René Chester Goduscheit, Doctoral Research Student
- Kristin Falck Saghaug, Ph.D. Fellow
- Carsten Bergenholtz, Research Assistant
- Andreas Slavensky, Doctoral Research Student
- Yariv Taran, Doctoral Research Student
- Peter Lindgren, P.h.D, Associate Professor

Aarhus School of Business, University of Aarhus
Haslegaardsvej 10, DK-8210 Århus V, Denmark
- Anders McIlquham-Schmidt, Ph.D., Assistant Professor
- Carsten Bergenholtz, Doctoral Research Student
- Kean Sørensen, Research Associate

University of Southern Denmark
Alsion 2, DK-6400 Sønderborg, Denmark
- Niels N. Grünbaum, Ph.D, Assistant Professor
- Svend Hollesen, Ph.D., Associate Professor

University of Southern Denmark
Campusvej 55, DK-5230, Odense M, Denmark
- Per Servais, Ph.D, Associate Professor
- Erik S. Rasmussen, PhD, Associate Professor

Contents

Figures

Tables

Chapter 1.0

Introduction

Peter Lindgren

The NEWGIBM project - has finally come to "an end of the beginning". The NEWGIBM project featured and made it possible for companies, researchers, consultancies and organizations together to analyze, conceptualize and even implement NEWGIBM – but it also "kick started" a whole new line of new projects – some would call these new BM.

The NEWGIBM project was original formed on the learning and results of 4 projects. The Ph.D. work on the PUIN project (Bohn and Lindgren 2002), The Network based High Speed Innovation project (Lindgren 2003), The EPUIN project (Bohn and Lindgren 2006) and the Born Global project (Madsen, Servais og Rasmussen 2002). On behalf of these 4 projects a group of 10 researchers from University of Southen Denmark, Aarhus School of Business at Aarhus University and Center of Industrial Production at the Aalborg University in Denmark sat down and formulated an application to the Danish Ministry of Science and Innovation.

The project became the name "The NEWGIBM project" addressing the need for SME´s to develop New Global business models supported by ICT and involving network partners to survive business in the future.

The NEWGIBM Project was accepted and funded in November 2005. The work was carried out in the time period until February 2008 – where a closing conference on NEWGIBM was held gathering national and international SME´s, organizations and researchers – amongst others keynote speakers

- Professor Adjunct Professor Henry Chesbrough, Executive Director, Center for Open Innovation Center for Open Innovation Management of Technology, University of California Berkley Hass School of Business
- Professor Christopher Tucci, Professor of Management of Technology at the Ecole Polytechnique Fédérale de Lausanne (EPFL).
- Alexander Osterwalder, École des Hautes Études Commerciales (HEC) (Business School)

This book which also serves as a part of the final evaluation and documentation of the NEWGIBM project describes the background, theory references, case studies, results and not least the learning about how to innovate BM. The New Global Business Model project supported by ICT gave us during the time period from 2005 – 2008 the platform to develop new business models as the following projects International Center for Innovation, M-commerce, Global Innovation, Global E-business, The Blue Ocean project, Intelligent Utility, Smart City, Women

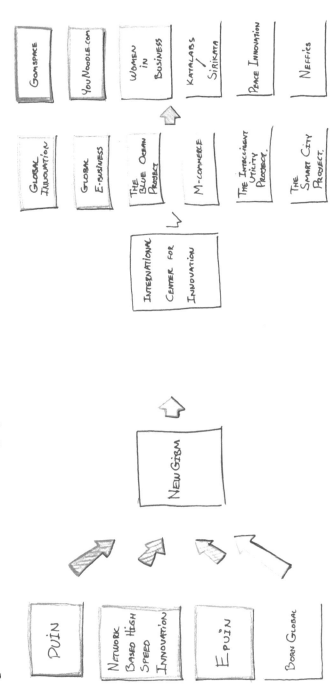

Figure 1: NEWGIBMs relations to other projects

in Business, Censec, The High Technology project – Gomspace, Younoodle.com project, The Katalabs project, The Neffics project and last but not least, The Peace Innovation project. The NEWGIBM project gave us a learning platform about how to innovate Global Business Models supported by ICT.

From research to local to global business model innovation

The book is a result of all this work, researchers and SME's efforts and collaborative work to understand Global Business Model Innovation. The potential that new business models create - please see some reference beneath - between different SME's, consultancies and researchers - across line of business, competencies and research domains - is enormous. The Books touch upon just some of these and what we really can do when we join forces.

The book describes and shows how the NEWGIBM project was carried out - as at the beginning a very much explorative analysis and work in 8 different industrial case companies/networks to later a more stable and focused research and business model development between partners.

New Business model innovation models and processes of NEWGIBM have been developed during the project. The researchers behind the project try by this book to describe and analyze the context, their experience and their scenarios of the NEWGIBM – project and beyond. This we hope will add new knowledge to new global business models based on ICT.

The book start off with an identification and introduction of the NEWGIBM challenges back in 2005 – what was the challenge behind NEWGIBM and development of NEWGIBM? - focusing on identification and introduction of why NEWGIBM was a challenge to SME companies?

The book endeavors to describe when NEWGIBM is a challenge and give an introduction to the theoretical model apparatus for NEWGIBM and the development processes and models behind. The book gives an introduction to when NEWGIBM models and processes can be used with preference.

The book gives an introduction to how the theoretical framework and research method behind the NEWGIBM project was outlined, developed and subsequently empirically tested and carried out in the case companies.

The book introduces some empirical showcases and studies of case companies, which have tried to establish an BM innovation platform for NEWGIBM processes and models.

The NEWGIBM project was right from the beginning developed and based on networks – because the researchers believed that this was the trend to BM innovation for the future. The SME´s tried to employ these models – some with success – others not. New global business model innovation is about trial and error.
The book finally explains how the study has contributed towards a deeper explanation to BM innovation processes and models of NEWGIBM.

On the basis of the above mention we would like on behalf of the whole research group behind the NEWGIBM project and ICI to thank the Ministry of Science and Innovation for funding the NEWGIBM project with 3,8 mio. DKr. The funding made it possible for us as researchers to begin our research on BM innovation together with some highly motivated case companies/networks – (Tricon, Danfoss, Elro, Smart House, Smart Habour, KMD, Skov A/S, Smart House). Due to confidentiality we were not allowed to publish the cases of Tricon, Elro and Skov A/S.

A big thank you to the SME´s for letting us in as researchers and opening up your companies and networks to us. As researchers we enjoyed and respect Your big and hard work on developing NEWGIBM. Without you as companies our study could not have happened.

Finally we would like to thank Sussanne, Christoffer and Allan for developing all the graphics to the book. Thank you Todd for translating and correcting our text.

A special thanks also to Helle for being our core point – collecting materials for chapters, arranging meetings and pushing us all to finish our work.

We would use this opportunity to welcome all interested to spend some hopefully enjoyable, interesting and learning hours together with the book – NEWGIBM.

Yours Sincerely

Peter Lindgren

www.globalinnovation.dk
www.globalebusiness.dk
www.blueocean.dk
www.m-commerce.dk
www.ici.aau.dk - International Center for Innovation
www.intelligentutility.dk - The Intelligent Utility project
www.smartcitydk.dk
www.womeninbusiness.dk
www.censec.dk
www.gomspace.com - High Technology project
www.katalabs.com
www.younoodle.com
www.neffics.eu
www.stanfordpeaceinnovationlab.org - The Peace Innovation project

References

- Bohn og Lindgren (2002), *Produktudvikling i Netværk*

- Bohn og Lindgren (2006), *Elektronisk Produktudvikling i Netværk - EPUIN projektet*

- Chesbrough et all (2008), *Open Business models*

- Lindgren (2003), *Network-based High Speed Innovation,* Ph.D. Dessertation, Aalborg University

- Lindgren, P., Taran, Y. & Saghaug (2010), *A Futuristic Outlook on Business Models and Business Model Innovation in a Future Green Society*

- Madsen T., Servais og Rasmussen (2002), *Born Global projektet*

- Magretta, J. (2002), *Why Business Models Matter,* Harvard Business Review 80(5): 86–92.

- Malhotra Yogesh (2000), *Knowledge management and new organization forms: A framework for business model innovation,* Information Resources Management Journal; 13, 1; ABI/INFORM Global pg. 5

- Osterwalder, Alexander (2004), *The business model ontology - a proposition in a design science approach,* University de Lausanne, Switzerland.

- Osterwalder, Alexander; Pigneur, Yves; Tucci L. Christopher (2004), *Clarifying Business Models: Origins, Presents, and Future of the Concept,* Communications of AIS, Volume 15, Article

- Pateli G. Adamantia and Giaglis M. George (2005), *Technology innovation-induced business model change: a contingency approach,* ELTRUN: Journal of Organizational Change Management Vol. 18 No. 2, pp. 167–183. The E-Business Centre, Athens University of Economics and Business, Athens, Greece

- Saghaug, K. M. & Lindgren, P. (2008), *Innovation Business models - a question of attracting different intellectual capabilities who can think differently*

- Tapscott, D., & Anthony D. Williams (2008), *Wikinomics,* ISBN 978-159184-193-7 Penguin Books LtD.

- Vervest et all (2005), *Smart Business Networks,* Van Nordstrand

- www.wikipedia.org/

Chapter I.1

The Theoretical History and Background of Business Models

Peter Lindgren
&
Yariv Taran

Abstract

The history of business model theory is considered to be relatively young, but the study has intensified in the late 90s and early 2000s. This chapter will explain the theoretical history and background of business models. The chapter describes four major inputs to the definition of business and business model - all of which have been built upon many other researchers' work.

Keywords: Business, Business Models, Innovation

Introduction

Business models are a challenge to innovators and effective business models are a tremendously valuable asset to a company (Chesbrough 2007). Most business leaders however, when asked to explain their company's business model, would not have a ready answer to give, and when they do come up with an answer, they would most likely present their organisational structure. But does this represent a holistic business model? Many managers do not really know what their business model represents. Do they even have an explicit business model? And assuming they do, do they know how to continually innovate it successfully?

What is it about business models that make them so difficult for leaders to comprehend? What is really known about the rationale of business models? What is the difference between the innovation of a product and the innovation of a business model?

According to Linder and Cantrell (2000), executives cannot even articulate their business models. Many leaders talk about business models but 99 percent have no clear framework for describing their own model. They do know what business they are in, they just cannot describe it clearly. And if they are unable to describe it clearly, they cannot share it effectively throughout their organisation. Many try to describe the business model but most are only touching "the elephant" (Chesbrough 2007).

Magretta (2002) continues in this line of thought and argues that both the terms 'business model' and 'strategy' are among the most sloppily used terms in business. 'They are often stretched to mean everything – and end up meaning nothing'. Nonetheless, according to her, these two concepts are of enormous practical value.

It seems as if there is a growing necessity to link together the rising gap between the intellectual environment, where authors with academic backgrounds have no trouble spinning theories on the contribution of various business models to organizations, and business leaders, who are more interested in finding proof that the theories are actually also effective in practice[1], and that they had a better chance of success by employing ideas from the business model theories. Overcoming this gap is, undoubtedly, the next true challenge.

At present there is extensive knowledge about innovation, in general, (Ulrich and Eppinger 2000, Tidd et al. 2005) and how to innovate products (Wind 1973, Cooper 1993, Baker and Hart 2007) in particular, but very little is known about how to innovate business models. The aim of this chapter, therefore, is to open the discussion on business model innovation as a new field of research. However, before discussing how to innovate business models, or when a business model could be defined as 'new', a definition, as well as its theoretical background, is required.

The History and Evolution of the Business Model (BM) Theory

What is a 'business' and a 'business model' (BM) really? Unfortunately the answer is inconclusive. Different authors will define the concept in dissimilar ways. In order to simplify things however, it would probably be easier to separate the question into two components: first the chapter begins by looking into how the concept of business has been defined, and later on it will clarify what is meant by the 'business model' definition and the functions of the business model.

Towards an Understanding of a Company's 'Business'

Derek F. Abell in his well known 1980 book: "Defining the Business", argued that a business could be defined according to a 3 dimensional framework:

1. Customer Functions: customer functions rendered by the company
2. Customer Groups - customer groups served by the company
3. Customer Technology - customer technology used to produce customer functions and serve customer groups

---

1   Woback P. James and Jones T. Daniel (2003)

Figure 1: A Three Dimensional Framework for the 'Business' Definition
Source: Derek F. Abell, 1980

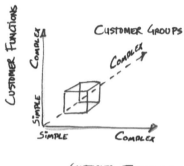

Abell argued that a company's core business, 'the strategic business unit (SBU)' is related to what is inside the box (as shown in the figure above). This is related to what the company had defined as its core business – the company's business model. Abell argued that it is particularly important to the company to know about and define the company's core business because of the strong relation of the core business to the company's mission, goals and strategy – the strategic perspective.

The SBUbox, as shown in the figure above, represents the core business of the company. The space external to the box represents the potential for innovation and further business development. Abell argued that a company, in general, should stick to its core business – core business model. By doing so, the company could achieve optimisation of business processes and resource utilisations.

When innovating out of the core business area, Abell advised the focal company to carefully analyse its own competences and resources. Abell argued that moving into a new strategic business field (the customer function, customer group or customer technology) would cause a demand for other resources and new competences and, thereby, call for innovation into the field of product development, market development or different degrees of diversification.

Abell's observations on core business have yielded 3 important dimensions on how to define business. They are building blocks of the business model. These building blocks are strongly related to the customer focus viewpoint – an 'outside-in' perspective - a strategic marketing perspective. The model is also based on a dyadic relationship between the focal company and the served customer group.

Abell's model explains very little about what the business model architecture really looks like, how it is being organised and how it functions. Although Abell's model operates with the 3 dimensions - which indeed covers and prioritises some of a company's BM and services towards customers' values and norms – it says nothing about what values the customer truly prioritises and perceives as values (Ravald & Grönross 1996). Neither does it say anything about costs and what the customer perceives as costs.

Furthermore, the customer groups in Abell's model focus solely on a downstream perspective, meaning that suppliers, and even the focal company itself, cannot be a customer in the business model. Both customer functions and customer groups, in Abell's model, can only be seen in a one dimensional scale. However, when both customer functions and customer groups are organised in chains of customer functions and groups, they become more interrelated and the model therefore has to be further developed. Where such characteristics exist, it becomes extremely difficult to identify the customer functions and customer groups and their interdependencies within the core business model. Also, the customer technology dimensions need to be detailed further. Finally, the financial aspects of the business are not really explained in Abell's model.

Nonetheless, Abell's model provided us a starting point towards understanding a company's core business logic. As the company's business environment grows more network oriented and based, more global, agile, and related to both upstream and downstream values, it offers us the challenge to better understand the overall BM concept.

Towards an Entrepreneurial Understanding of Business Models

As mentioned earlier, the history of business model and business model innovation theory is relatively young. However, the study has intensified lately. Morris, Schindelhutte and Allen (2003) presented another view on business models, which was a more interrelated BM framework with a strong focus on an entrepreneurial understanding of business models. They tried to build what they called 'a unified perspective of business models' on top of academic work that had been carried out since the late nineties and up to 2003. This can be seen in the table below.

Table 1: Perspectives on Business Model Components (Morris et al. 2003)

Source	Specific Components	Number	E-Commerce/ General	Empirical Support (Y/N)	Nature of Data
Horowitz (1996)	Price, Product, Distribution, Organisational Characteristics and Technology	5	G	N	
Viscio and Pasternak (1996)	Global Core, Governance, Business Units, Services and linkages	5	G	N	
Timmers (1998)	Product/Service/Information Flow Architecture/Business Actors and Roles/Actor Benefits/Revenue Sources/ Marketing Strategy	5	E	Y	Detailed Case Studies
Markides (1999)	Product Innovation, Customer Relationship, Infrastructure Management, Financial Aspects	4	G	N	
Donath (1999)	Customer Understanding, Marketing Tactics, Corporate Governance, Intranet/ Extranet Capabilities	5	E	N	
Gordijn et al. (2001)	Actors, Market Segments, Value Offering, Value Activity, Stakeholder Network, Value Interfaces, Value Ports, Value Exchanges	8	E	N	

Source	Specific Components	Number	E-Commerce/ General	Empirical Support (Y/N)	Nature of Data
Linder and Cantrell (2001)	Pricing Model, Revenue Model, Channel Model, Commerce Process Model, Internet-Enabled Commerce Relationship, Organisational Form, Value Proposition	8	G	Y	70 Interviews with CEOs
Chesbrough and Rosenbaum (2000)	Value Proposition, Target Markets, Internal Value Chain Structure, Cost Structure and Profit Model, Value Network, Competitive Strategy	6	G	Y	35 Case Studies
Gartner (2003)	Market Offering, Competencies, Core Technology Investments, Bottom Line	4	E	N	Consulting Clients
Hamel (2001)	Core Strategy, Strategic Resources, Value Network, Customer Interface	4	G	N	Consulting Clients
Petrovic et al. (2001)	Value Model, Resource Model, Production Model, Customer Relation Model, Revenue Model, Capital Model, Market Model	7	E	N	
Dubosson-Torbay et al. (2001)	Products, Customer Relationship, Infrastructure and Network og Partners, Financial Aspects	4	E	Y	Detailed Case Studies

Source	Specific Components	Number	E-Commerce/ General	Empirical Support (Y/N)	Nature of Data
Afuah and Tucci (2001)	Customer Value, Scope, Price, Revenue, Connected Activities, Implementation, Capabilities, Sustainability	8	E	N	
Weill and Vitale (2001)	Strategic Objectives, Value Proposition, Revenue Sources, Success Factors, Channels, Core Competencies, Customer Segments, IT-Infrastructure	8	E	Y	Survey Research
Applegate (2001)	Concept, Capabilities, Value	3	G	N	
Amot and Zott (2001)	Transaction Content, Transaction Structure, Transaction Governance	4	E	Y	59 case Studies
Alt and Zimmermann (2001)	Mission Structure, Processesses, Revenues, Legalities, Technology	6	E	N	Literature Synthetis
Rayport and Jaworski (2001)	Value Cluster, Market Space Offering, Resource System, Financial Model	4	E	Y	100 Cases
Betz (2002)	Resources, sales, profits, Capital	4	G	N	

Morris *et al.* have concluded that the findings of their cross-theoretical perspective (table 1 above) have led them to believe that no single theory can fully explain the value creation potential of a business enterprise. Consequently, it is impossible to identify a holistic building block framework for a generic business model depending merely on one author's perspective. They could identify, however, common central ideas and theories with the following details and aspects:

Business Models:

- Build upon central ideas in business strategy and theoretical traditions
- Relate to the focal firm's economic model
- Build upon the value chain concept (Porter, 1985)
- Extend the notion of value systems, strategy positioning, and competitive advantage (Porter, 1996)
- Are described by the resource based theory which relates the firm to a network (Barney et al. 2001)
- Relate to the strategic network theory (Jarillo, 1995) and to cooperative strategies (Dyer and Singh, 1998)
- Involve choices concerning a firm's boundaries - vertical integration (Barney, 1999), transaction cost economics (Williamson, 1981)
- Include the firm's services and products (customer functions) and activities undertaken to produce them (customer technology) (Abell, 1983)
- Are the firm's architectural backbone

Morris et al. continue by arguing that a business model framework must be reasonably simple, logical, measurable, comprehensive, operational and meaningful. The challenge, therefore, is to produce a framework that is applicable to firms, in general, but also serves the needs of the individual entrepreneur. The focus of Morris et al. is strongly related to the focal firm and the individual entrepreneur.

As a result of their findings, Morris et al. propose a BM framework with three specific decision making levels:

1. **Foundation:** Defining basic components. Need to make generic prioritisation decisions concerning what the business is and is not, as well as ensuring internal consistency of decisions
2. **Proprietary:** Development of marketplace advantage through unique decision combinations
3. **Rules:** Providing specific governing guidelines for future business operations improvements

In addition to these three guidelines, they have developed an interactive framework that includes the three specific decision making levels mentioned above with relation to six basic 'decision areas' of considerations. These are shown in the table below.

Table 2: Morris et al. BM Framework Proposal

Basic Decision Areas	Levels of Decision Making Areas			
	Question	Foundation	Proprietary	Rules
Component 1: Factors Related to services and products	How do we create value? Primarily products/primarily services/heavy mix Standardised/some customisation/ high customisation Broad line/medium breadth/narrow line Deep lines/medium depth/shallow lines Access to product/ product itself/ product bundled with other firm's product Internal manufacturing or service delivery/ outsourcing/licensing/ reselling/ value added reselling Direct distribution/indirect distribution (if indirect: single or multi channel)			
Component 2: Market Factors	Who do we create value for? Type of organization: b-to-b/b-to-c/ both Local/regional/national/international Where customer is in value chain: upstream supplier/ downstream supplier/ government/ institutional/ wholesaler/ retailer/ service provider/ final consumer Broad or general market/multiple segment/niche market Transactional/relational			
Component 3: Internal Capability Factors	**What is our source of competence?** Production/operating systems Selling/marketing Information management/ mining/packaging Technology/R&D/creative or innovative capability/intellectual Financial transactions/arbitrage Supply chain management Networking/resource leveraging			

Basic Decision Areas	Levels of Decision Making Areas			
	Question	Foundation	Proprietary	Rules
Component 4: Competitive Strategy Focus	**How do we competitively position ourselves?** Image of operational excellence/ consistency/dependability/speed Product or service quality/ selection/features/availability Innovation leadership Low cost/efficiency Intimate customer relationship/experience			
Component 5: Economic Factor	How do we make money? Pricing and revenue sources: fixed/mixed/flexible Operating leverage: high/medium/low Volumes: high/medium/low Margins: high/medium/low			
Component 6: Growth/ Exit Factor	What are our time, scope and size ambitions? Subsistence model Income model Growth model Speculative model			

Morris et al. argued that the three decision levels – foundation, property and rules – reflect the different managerial purposes of the business model. The proposed framework provides users with the possibility to design, describe, categorise, criticise and analyse their business model – for any type of company. They also claimed that a considerable scope of innovation potential exists in each model component, and that the holistic framework model suggested here can open a path towards new and general business model taxonomies and archetypes. They did not mention, however, any archetype business model framework explicitly. Nonetheless, they still succeeded in tremendously improving our understanding of the contour of business models – yet still from an entrepreneurial business model view.

Towards Building Blocks of a Business Model

Corresponding with the findings of Morris et al., Osterwalder, Pigneur and Tucci (2004) have also developed a framework for business models. They have summed up the academic work of previous business models adding new theoretical aspects. Their research went back more years than Morris et al. (the last 20 years) and continued up until 2007. They defined the BM as 'A blueprint of how a company does business'. It is 'a conceptual tool that contains a set of elements and their relationships and allows expressing a company's logic of earning money. It is a description of the value a company offers to one or several segments of customers and the architecture of the firm and its network of partners for creating, marketing and delivering this value and relationship capital, in order to generate a profitable and sustainable revenue stream'.

According to this definition, business models serve as a platform that represents a company's operational and physical manifestation. Thus, the challenge for a business model 'designer', before innovating the model, is to first identify the key elements and the key relationships that describe a company's 'AS-IS' business model before innovating it.

This definition was designed upon five phases of evolution within the business model literature:

Figure 2: Evolution of the Business Model Concept
(Source: Osterwalder, Pigneur and Tucci 2004)

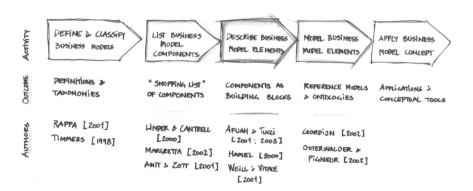

One of the most interesting points suggested here, is that at present business models are accepted as being constructed not only with visual, but also with conceptual components – 'meta-models in the form of reference models and ontology' (Gordijn 2002; Osterwalder & Pigneur 2004). The figure above also shows that innovating BMs is a new field of exploration of BMs.

Osterwalder, Pigneur and Tucci have also tried to create a common language when addressing business model characteristics. The following table presents 'nine building blocks' components for the BM's framework. The table suggested here, is a result of a study and a research summation of fourteen different authors[2], who attempted to identify different components of BMs.

Table 3: The Nine Building Blocks of a Business Model
Source: Lindgren and Taran based on Osterwalder et al[3]. (2004); Morris et al (2003); and Abell (1983)

Product	Value Proposition	Gives an overall view of a company's bundle of products and services	Customer Function	Component 1: Services and products
Customer Interface	Target Customer	Describes the segments of customers a company wants to offer value to	Customer Group	Component 2: Market Factors
	Distribution Channel	Describes the company's various means of getting in touch with its customers	Customer Technology	Component 3: Internal Capability Factors
	Relationship	Explains the kind of links a company establishes between itself and its different customer segments	New	Component 3: Internal Capability Factors

2   Stähler 2001; Weill et al. 2002; Petrovic, Kittl et al 2001; Gordijn 2002; Afuah and Tucci 2003; Tapscott, Ticoll et al. 2000; Linder and Cantrell 2000; Hamel 2000; Mahadevan 2000; Chesbrough and Rosenbloom 2000; Magretta 2002; Amit and Zott 2001; Applegate 2001; Maitland and Van de Kar 2002

3   The original Table of Osterwalder et al. has been written in bold letters in Table 3

Infrastructure Management	Value Configuration	Describes the arrangement of activities and resources	Customer Technology	Component 3: Internal Capability Factors
	Core Competence	Outlines the competences necessary to execute the company's business model	New	Component 3: Internal capability factors
	Partner Network	Portrays the network of cooperative agreements with other companies necessary to efficiently offer and commercialize value	New	New (Maybe - Component 3: Internal Capability Factors)
Financial Aspects	Cost Structure	Sums up the monetary consequences of the means employed in the business model	New	Component 5: Economic Factors
	Revenue Model	Describes the way a company makes money through a variety of revenue flows	New	Component 5: Economic factors
Strategy aspects	Strategy model	Not Existent	Not Existent	Component 4: Competitive Factor and Component 6: Personal/ Investor Factor

As can be seen, Osterwalder et al. overlap on 4 areas with Abell but have added 5 new elements to our understanding of the focal company's business model – Relationship, Core Competence, Partner Network, Cost Structure, and Revenue Model. Both Abell and Osterwalder at al. lack the strategy dimensions related to Morris at al.

Towards an Open Business Model

Chesbrough (2007) introduced a whole new way of thinking about business models and innovation of business models in his book *Open Business models – How to Thrive in the New Innovation Landscape.* Firstly, he argued towards open innovation (2005) and then opening companies' business models (2007). Until then, all academic work (Abell 1983, Linder and Cantell 2000, Morris 2003, Magretta 2000, Osterwalder et all 2004) and practically all business thinking had been related to a 'closed' business model – sticking to your core business – where the business model was strongly related to a single, focal company. Chesbrough argued that 'a closed business model' was not efficient for future innovation and a global competitive environment. Much innovation was never even related to the core business model and companies who could have used the results of the innovation were prevented from this because of patents, unwillingness to network and even by lack of knowledge by those who had innovated.

Chesbrough argued that, in the future, companies would have to open up their business models and allow other companies to integrate them with their own business models and even take parts out of the model to use it in other business models. These innovations could be extremely valuable and more effectively used in other business models by other companies.

Because of 'a new global business environment' the business models of companies have to be structured and managed in a more open way – and e.g. developed and registered patents should become open to other companies of interest in order to support and develop their business models or the innovation of new business models. These innovations could become value adding building blocks to both existing and new business models.

This is the case because companies are playing by different rules using different business models. An open approach will give new opportunities and challenges for both 'giver' and 'taker' of the innovation. It will completely restructure the theory of business models leading to a much more network oriented business model. New business models will focus more on different partners' value equation and will be more open to network partners, both within and outside the value chain.

The open innovation model has become an important concept for further work and development of business models.

Reflection and Conclusion

Business models have, until now, basically been stories explaining how enterprises work. According to Magretta (2002), a good business model answers questions like: who is the customer (customer groups)? and what is the customer value (customer functions)? It also answers the fundamental questions every manager must ask: how do we make money in this business (the revenue model)? and what is the underlying economic logic that explains how we deliver value to customers (customer technology) at the appropriate cost?

A new understanding and theory of business models and what they are all about seems to be developing quickly at present. A more network based business model moving from a closed to open is developing. It is much more focused on values and cost, not only of partners inside the business models but also those outside – partners who are either in the first, second or even other levels of the chain of customers. In the figure below, all the main findings and contributions to the science about business models are shown.

Figure 3: Main contributions to the research on BM from 1980 - 2007 [4]

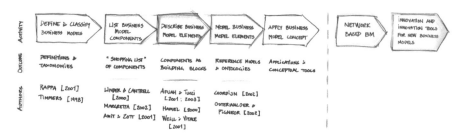

These elements and findings were an important starting point for the NEWGIBM research group to continue the research on defining, developing and innovating 'New Global ICT Business Models'.

The next chapter will continue to explore such questions and will extend its comments on the views and issues raised by the existing literature mentioned above, particularly with the intention of examining when a BM can be defined as being new.

---

4   The first five phases are based on the findings of Osterwalder *et al.*

References

- Abell, D.F. (1980), *Defining the Business: The Starting Point of Strategic Planning*, Prentice-Hall, Inc. N.J.

- Afuah, A. and C. Tucci (2003), *Internet Business Models and Strategies,* Boston: McGraw Hill.

- Alt, R. and H.D. Zimmerman (2001), *Introduction to Special Section on Business Models.* Electron Mark, 11(1): pp. 3–9.

- Amit, R. and C. Zott (2001), *Value creation in e-Business, Strategic Management Journal*, Vol. 22, Nos. 6–7, pp. 493–520.

- Applegate, L. M. (2001), *E-business Models: Making Sense of the Internet Business Landscape.* In G. Dickson, W. Gary and G. DeSanctis, *Information Technology and the Future Enterprise: New Models for Managers*, Upper Saddle River: Prentice Hall.

- Baker, M. and S. Hart (2007), *Product Strategy and Management*, Harlow, U.K., Prentice Hall, pp 157–196.

- Barney, J.B. (1999), *How Firm Capabilities Affect Boundary Decisions, Sloan Manage Rev*, 40(1): pp.19–32.

- Barney, J., M. Wright, and D. Ketchen (2001), *The Resource-Based View of the Firm: Ten Years After, J Manage*, 27(6), pp.625–41

- Chesbrough, H. and R. S. Rosenbloom (2000), *The Role of the Business Model in Capturing Value from Innovation: Evidence from XEROX Corporation's Technology Spinoff Companies'*, Boston, Harvard Business School Press.

- Chesborough, H (2005) Open *Innovation - The new Imperative for Creating and Profiting from Technology* Harvard Business School, Press

- Chesborough, H (2007) *Open Business Models How to Thrive in the New Innovation Landscape* Harvard Business School ISBN 13: 978-1-422-1-0427-9

- Cooper, R. (1993), *Winning at New Products: Accelerating the Process from Idea to Launch'*, Reading: Addison-Wesley.

- Dubosson -Torbay M., A. Osterwalder, and Y. Pigneur (2001), *E-business Model Design, Cclassification and Measurements,* Thunderbird Int. Bus Rev. 44(1): pp. 5–23

- Donath, R. (1999), *Taming e-Business Models*. ISBM Business Marketing Web Consortium 3 (1). State College (PA): Institute for the Study of Business Markets; pp. 1–24.

- Dyer, J and H. Singh (1998), *The Relational View: Cooperative Strategy and Sources of Interorganizational Competitive Advantage*, acad. Manage. 23(4), pp. 79–660.

- Gartner, (2003), Available at: http://www3.gartner.com, 2003.

- Gordijn, J. (2002), *Value-based Requirements Engineering - Exploring Innovative E-Commerce Ideas*, PhD thesis, Vrije Universiteit Amsterdam, The Netherlands.

- Hamel, G. (2000), *Leading the Revolution*, Boston: Harvard Business School Press.

- Horowitz, A.S. (1996), *The Real Value of VARS: Resellers Lead a Movement to a New Service and Support*, Mark Comput; 16, (4): pp. 6–31

- Jarillo, J. C. (1995), *Strategic Networks*. Oxford: Butterworth-Heinemann.

- Linder, J. and S. Cantrell (2000), *Changing Business Models: Surfing the Landscape*, Accenture Institute for Strategic Change, Canada.

- Lindgren, P. (2003), *Network Based High Speed Product Innovation*, PhD thesis, Centre for Industrial Production, Aalborg University, Denmark.

- Lindgreen, A. and F. Wynstra (2005), *Value in Business Markets: What Do We Know? Where Are We Going? Industrial Marketing Management*, Vol. 34, Issue 7, pp. 732–748.

- Magretta, J. (2002), *Why Business Models Matter? Harvard Business Review*, Vol. 80, No. 5, pp. 86–92.

- Mahadevan, B. (2000), *Business Models for Internet-based e-Commerce: An Anatomy, California Management Review*, Vol. 42, No. 4, pp. 55–69.

- Markides, C. A., (1999), *Dynamic View of Strategy*, Sloan Manage Rev; 40(3): pp. 55–63.

- Morris, M, M. Schmindehutte and J. Allen (2003), *The Entrepreneur's Business Model: Toward a Unified Perspective, Journal of Business Research*, 58, 6, pp. 726–735.

- Osterwalder, A., Y. Pigneur and L.C. Tucci (2004), Clarifying Business Mmodels: Origins, Present and Future of the Concept, Communications of AIS, No. 16, pp. 1–25.

- Osterwalder, A. and Y. Pigneur (2004), *An Ontology for e-Business Models*, University of Lausanne, Switzerland.

- Petrovic, O., C. Kittl and R.D. Teksten (2001), *Developing Business Models for e-Business*, Proceedings of the International Conference on Electronic Commerce, Vienna.

- Porter, M.E. (1985), *Competitive Advantage,* New York: Free Press

- Porter, M.E. (1996), *What is Strategy?,* Harvard Bus Rev, 74(6), pp.61–78

- Rappa, M. (2004) *The Utility Business Model and the Future of Computing Services, IBM Systems Journal* 43(1): pp. 32–43

- Ravald, A., and C. Grönroos (1996), *The Value Concept and Relationship Marketing, European Journal of Marketing*, Vol. 30 No.2, pp.19–30.

- Rayport, J.F, and Jaworski, B.J (2001), *E-Commerce*, NewYork: McGraw-Hill/Irwin.

- Stähler, P. (2001), *Geschäftsmodelle in der Digitalen Ökonomie. Merkmale, Strategien und Auswirkungen*, PhD thesis, University of St. Gallen, Switzerland.

- Tapscott, D., D. Ticoll, A, Lowy (2000), *Digital Capital. Harnessing the Power of Business Webs*, Boston: Harvard Business School Press.

- Tidd, J., J. Bessant, K. Pavitt (2005), *Managing Innovation. Integrating Technological, Market and Organizational Change*, Chicester: John Wiley & Sons.

- Timmers, P. (1998), *Business Models for Electronic Markets, Journal on Electronic Markets* 8(2): pp. 3–8.

- Ulrich, K.T and S.D. Eppinger (2000), *Product Design and Development*, USA: Irwin McGraw-Hill.

- Viscio, A. J. and B.A. Pasternack (1996), *Toward a New Business Model,* Strateg Bus 2(1): pp. 125–34.

- Weill, P. and Vitale, M.R. (2001), *Place to Space*, Boston: Harvard Business School Press.

- Weill, P., M. Subramani and M. Broadbent (2002), *Building IT Infrastructure for Strategic Agility, Sloan Management Review*, Vol. 44, No. 1, pp. 57–65.

- Wind, Y. (1973) *A New Procedure for Concept Evaluation, Journal of Marketing*, Vol. 37, October, pp. 2–11.

- Williamson, O.E. (1981) *The Economics of Organization: the Transaction Cost Approach*, Am J Sociol 87(4), pp. 77–548.

- Woback, P. James and Daniel T. Jones (2003) *Lean Thinking: Banish Waste and Create Wealth in Your Corporation*, N.Y.: Free Press

Chapter I.2

The Theoretical Background of Business Model Innovation

Peter Lindgren
&
Yariv Taran

Abstract

Innovation today is changing from innovating excellent products, processes or services to innovating new business models. Innovating a holistic, sustainable and open business model - an outside-in business model - which includes both intra-firm as well as collaboration and networking capabilities is not widely explored. Consequently, this topic is in its embryonic stage, both theoretically and in managerial practice. This chapter examines the development of business models innovation and focuses on the challenges related to that.

The aim of this chapter is to answer the research question, namely; what is a new business model?

Keywords: Innovation; New Business Models; Cross Functional Innovation

Introduction

In today's complex knowledge and innovation driven economy, innovating business models and their architecture are in growing focus and demand. Innovating the business model to become open and network oriented is, however, complex. Yet this is critical for a company's future survival (Chesbrough 2007), particularly due to the fact that it provides a new and often unknown dimension to the innovation process of the firm.

Business leaders today have well realised, in the past five to ten years, that innovation based solely on product innovation, and aimed towards serving local markets alone is not sufficient in sustaining the competitiveness and survival of their companies. Competitors can easily and quickly copy product innovation and local market segments today are often quickly captured by global rivals located outside of the focal country.

Opposed to product innovation, business model innovation is proposed as a new line of innovation and a new way to meet the competition on global innovation. Business model innovation is considered much more difficult for rivals to copy and sometimes rivals even work together and cooperate on the innovation of the new business model. Therefore, innovation of new business models can be regarded as a long term sustainable asset to the company contrary to product innovation. This is not to say that companies should not work on product, process and other known innovation activities (Tidd 2005). They should, however, move their innovation focus up to a higher strategic level by innovating new business models.

Consequently, the great challenge of leaders today is, therefore, to choose which innovation activities to carry out and to know how to manage business model innovation in practice. That will secure their continuous innovation capabilities in the long run.

The aim of this chapter is, therefore, to answer the research question: what is a new business model?

The Nature of a 'New' Business Model

An interesting question arises, when one tries to determine if a business model should be characterised as new, namely: to whom should it be new? Is it enough that one of the participants in the network thinks it is new, or should all the participants have to agree that it is new before it is? Is it new if it is new to the participants but not generally within the industry as such? Is it a new business model if it is used in another industry but not in the one that the participants are a part of? Or should it be completely new to all, like e.g. the Internet and online shopping in the mid nineties?

The Level of Incremental and Radical Innovation

This discussion can, firstly, be related to the debate on defining incremental and radical product innovation (Rosenau 1993, Leifers 2002, Tidd et al. 2005).

Figure 1: Incremental and Radical Product Development
Source: Lindgren & Bohn, 2002

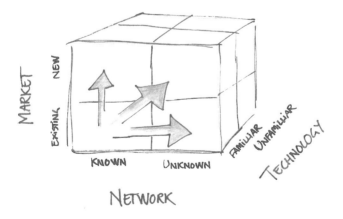

When an innovation focus on unfamiliar markets includes unfamiliar technology and cooperation with unfamiliar networks, it can be classified as radical innovation and as new. Among these characteristics will be different degrees of radicalism of the innovation. This model primarily focuses on characteristics outside the company – new and unknown markets, bringing in new and unknown technology and cooperating with new and unknown networks to characterize when an innovation project can be classified as radical.

Another way to classify whether an innovation project is incremental or radical is by using Tidd, Pavitt and Bessant's model (Tidd 2005). According to Tidd et al. innovation is about change and such innovation can touch upon 4 innovation areas each out of which can be characterised as incremental or radical:

- **Product Innovation:** concerns change in the products/services of the company
- **Process Innovation:** concerns change in the way products/services are created
- **Position Innovation:** concerns change in the context/market, on which the products/services of the company are introduced
- **Paradigm Innovation:** concerns the change in the mental models which serve as guidelines for the actions of the company

Figure 2: Incremental and Radical Product Development
Source: Tidd, Pavitt and Bessant, 2005

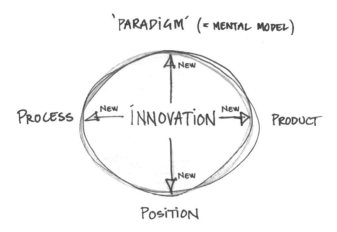

The complexity or radicality of the innovation can be considered as small or large changes within each of the four domains of the model or a combination which can be characterised as:

- 'Do better' innovation or 'improvement' innovation
- 'Do different' innovation

Incremental innovation is, therefore, characterized as mainly using existing knowledge to improve existing products/services. It is short term based and it focuses on the focal company performing better than it already does.

In opposition to this, radical innovation is based on the ability to find new knowledge and new possibilities. The innovation method is based on competences, that the company does not possess and can result in products/services that cannabalise or remove the basis of existing products/services.

A business model, however, consists of more building blocks than those of products, process, position and paradigm – as can be seen in the previous chapter (Table 3). The above mentioned innovation view explains only one small part of the story. The discussion, therefore, needs to move on to a strategic level.

The Level of Diversification

In general, business models are static by nature. They all merely provide a snapshot of the current situation of the company (Osterwalder 2004). Every business model is continually changing through time – particularly under the innovation process. Therefore, it is essential to introduce the concept of change and time when addressing new business models.

Abell (1980) argued that a company wanting to innovate had the potential to increase or decrease the business along 3 dimensions.

Abell related this to the company's choice of strategy for innovation. Innovating with small corrections on the 3 dimensions could be classified as an incremental innovation (Balacandra 1983). This type of innovation would result in:

1. Product Innovation: more or fewer customer functions
2. Market Penetration or Market Development: more customer groups in existing markets

3. Internal Innovation in the company: new production systems, new ways of organising the company resources

Figure 3: Innovating the Core Business
Source: Derek F. Abell, 1980

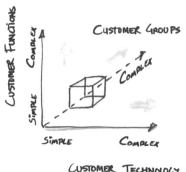

Most of this innovation could be classified as incremental innovation. On the contrary, innovating into completely new customer functions, new customer groups and new customer technology could be classified as radical innovation (Leifers 2002). Leifers defined this as a new product or a new technology for the market. Applying the Blue Ocean Strategy Theory (Chan and Mauborgne 2005) this would include moving into a market without competitors or where competitors are marginalised.

Abell's theorem of innovating radically on all thre dimensions at the same time could be defined as a totally new business model. Such innovation would include completely new costumer functions, new customer groups and new customer technology. Such innovation, happening concurrently would abandon the old business and move the company into a new strategic business unit (SBU). This can be defined as conglomeratic diversification (Kotler 2006) when moving on all three dimensions and could be defined as a truly new business model.

This type of innovation could be defined as a new core business of the company – a new strategic business unit (SBU) – and, therefore, be classified as innovating a completely new business model.

As many have already verified (Wind 1973, Eppinger 2004), this is very seldom done and considered to be highly risky. Normally, innovation, therefore, deals with small innovations and variations of a new business model – incremental innovation. The level of newness to the BMs change is related to the level of its diversification – the value proposition and configuration chosen. Below, three types of diversification are shown:

Figure 4: Innovating a New Strategic Business Unit
Inspired by Derek F. Abell, 1983

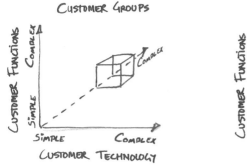
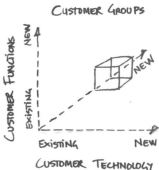

A genuinely new business or new business model is, in this terminology, equal to dealing with a conglomeratic diversification. The new business has until now been understood as highly risky and the chances for success are very small (Wind 1973, Kotler 1986, Leifers 2002). The reason for such a conclusion that it moves the company's business into a new business model and therefore is different and

Table 1: Different Types of Diversification
Lindgren 2007 based on Derek Abel 1980 and Kotler 1985 and 2006

Concentric Diversification	New products (functions) or new customer groups with technological and/or marketing synergies with existing product lines and customer groups – may appeal to a new class of customers or customer group
Horizontal Diversification	New products (functions), new customers (groups) with new unrelated technology unrelated to the current product line or current customer group – may appeal to a new class of customers or customer group
Conglomeratic Diversification	New business that has no relationship to the company's current customer functions (products), customer groups (markets) or customer technology

outside the focal company's core business and existing model. It changes the focal company's business model completely. This mindset of high risk with a low chance of success is changing today.

Abell argues that keeping only one or two of the dimensions in his model intact is less risky. That will increase the chances of success tremendously as opposed to change on all 3 dimensions. It has been verified, however, that this line of innovation is often more risky and related to lower chances of success (Christensen 2005, Chesbrough 2007).

All 3 variations of diversification can be considered as new business and as a new business model. Conglomeratic diversification is a truly new business model. As mentioned earlier, the network dimension is an important parameter in determining whether a business model is new. It is necessary to look carefully into the network dimension to understand better what is new in a new business model.

The Level of Change in Network

Many industrial companies have seen the necessity, in recent years, of applying network based innovation in order to optimise and to speed up innovation in order to compete in the global economy. Previously, companies "simply" had to match their product development competences with the market and the technology, but today, a match with the network component is vital. The competition in the 21st century is not company against company, but network against network. This is because a network no longer merely is physical and stable but encompasses all types of networks including physical, ICT, dynamic and virtual networks, as predicted before (Coldmann & Price, 1996, Child & Faulkner 1998, Vervest et al. 2006).

The consideration of when a network is new is not about whether a network is physical, digital or virtual. It is about whether the company collaborates in networks and if the focal company is familiar with such a collaboration. Research shows that new business models cannot be developed without entering a network. Neither can they be new if they do not enter continually into new types of networks with new partners. This is the case because the networks hold the knowledge to facilitate the innovation and when put together in the right constellation enable the innovation of new business models. Old networks will often only develop incremental and improved business models (Chesbrough 2007).

This also addresses the question of competences. If the innovation of business models is based only on competences within the company or within some old familiar networks or line of business, the business model will seldom be classified as new. Cross functionality and dynamic and flexible networks allow for new business models and thrust the business into new and undiscovered business areas (Chesbrough 2007).

There remain a number of dimensions to consider in order to define the new business as new. Therefore, other dimensions must be considered in order to understand better what is new.

The Level of Change in the Business Model

Linder and Cantrell (2000) argued that in order to straighten out the confusion of what a 'new' business model is, it is important to discuss new business models (or 'change models', in their terminology) based on four basic approaches: realisation models, renewal models, extension models and journey models.

'Realisation models', which cover most companies today, focus primarily on the exploitation of a current potential within an existing operational framework. This 'new' business model is considered to be the one with the least actual change. Examples of this are: geographical expansion of the firm, slight changes within the product line and continuously improving customer service. 'Renewal models' are firms that leverage their core skills to create a, sometimes disruptive, new position on the price/value curve. Linder and Cantrell use examples such as: revitalisations of product and service platforms, brand, cost structures and technology bases. Changes at this level can be addressed as 'blue ocean' (Chan and Mauborgne 2005) changes, achieved by attacking untouched markets as well as 'value innovation' efforts (increasing value and reducing costs). 'Extension models' include radical changes by finding new markets, changing value chain functions and improving product and service lines. This kind of model frequently involves forward and/or backward integration. The changes in the business model framework at this level are considered to be radical, mainly because of the fact that they are 'leveraging the firm's internal capabilities to create a complete new line of business' (Linder and Cantrell 2000). Finally, the 'journey models' can be considered as a complete transformation of the original business model. Here the company moves deliberately and purposefully toward a new operating model.

Figure 5: Change Models
(Source: Linder and Cantrell 2000)

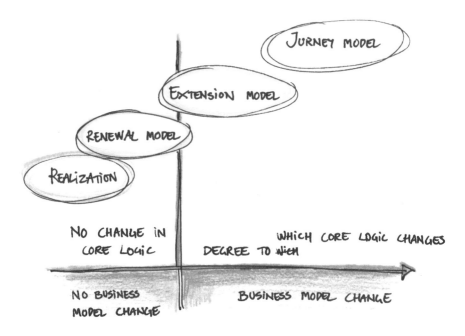

It is difficult to use the term 'new' business model when addressing realisation models since there is not any real change within the core logic (or operational business model) of the firm. Renewal and extension models can both be understood as an example of new business models, with degree of newness depending on the level of change within the company's operational and strategic functions. Finally, the journey model can clearly be regarded as leading to radical change.

The Level of Openness in Innovation

The aspects of 'closedness' and 'openness' of a business model were commented on earlier. Then the use of unfamiliar networks to innovate new business models was related to those aspects. Chesbrough comments on the need of open innovation (Chesbrough 2005) stressing this as a vital component in the innovation of new models. The authors take this argument further by claiming that a new business model must be an open model (Chesbrough 2007). A new business model cannot be new if it is not open because it will not continually be able to open new markets nor integrate with new technologies and new networks. The business model will quickly vanish from the market because it will be outdated soon after it has entered the market. It is not sustainable and built for continual innovation.

Later it will be discussed how this relates to the technical model or method. This is also related to the technical model or method of innovation as a support function to the new business model. An innovation model that is flexible (Creinert 1996, Verganti 1999, Corso 2004), open and dynamic can support a new business model.

Reflection

On the basis of Abell (1983), Kotler (1985), Linder and Cantrell (2000), and Chesbrough (2005, 2007) it can be seen that innovating new business models contributes to redefining a platform and new architectures for sustainable and long term innovation, as can be seen from table 2 below.

The global and the ICT aspects of the new business model will be addressed in one of the following chapters.

Table 2: Characteristics of New Business Models

Characteristics	'Old' Business Model *Formation*	A New Business Model
1. Innovation	Incremental product, process, position and paradigm innovation	Radical product, process, position and paradigm innovation
2. Markets		
Customer Functions	Familiar customer functions	Unfamiliar customer functions
Customer Group	Familiar customer group	Unfamiliar customer group
Customer Technology	Familiar customer technology	Unfamiliar customer technology
3. Diversification	Product and market development Integrative growth	Diversification

Characteristics	'Old' Business Model *Formation*	A New Business Model
3. Network	No networks or familiar networks either physical/ digital/ virtual Networks	Unfamiliar networks either physical/ digital/ virtual networks
4. Competences	Based solely on internal competence and familiar competences	Based on cross functional competences - across networks
5. BM	Individual, closed	Network-based, open
6. Success Criteria	Short term	Long terrm
7. Innovation Technique	Innovation based mainly on incremental product innovation, process innovation	Innovation based on business platform innovation with the possibility to innovate a variety of products/services, processes e.g. on the platform. Open innovation
8. Global	Innovation and the business model based on national market	Innovation and business model based on the global markets

Innovation studies today focus mainly on the characteristics and behavior of optimising a company's innovation activities to mostly short term success criteria (performance, cost, time) and more seldom on long-term success criteria (continual improvement, continual innovation and learning). Innovating new business models instead concerns and relies on cross-functionality, where highly skilled people tie several types of business development expertise together. They also create synergies, design winning new business models and sustain the new business system leading people and companies to put their competences and plans into action.

The success criteria of the new business model must be related to long term success criteria such as continual improvement and innovation and learning (Bohn & Lindgren 2005).

Furthermore, the new business model has a much stronger focus on how innovation ultimately creates value and diminishes cost to customers, suppliers and other participants in the innovation process. It focuses on the creation of both value and cost but also on perceived value and perceived cost. It will be commented further on later.

Later the considerations underlying the gathering of empirical qualitative data will be presented which constitute a very important part of the NEWGIBM project as well as reflections on how to handle qualitative data.

Conclusion

Today, company leaders are beginning to realise that the source of sustainable competitive advantage is changing from an excellent product, process or service innovation performance to having a more holistic, sustainable business model, which includes both intra-firm activities as well as collaboration and networking capabilities. This transformation is mainly due to the fact that product, process or service innovation, to a large extent, can easily be imitated by competitors, while business models are much more difficult to copy, if at all. Therefore, the main effort today is to know how to best lead organisational, distinctive innovation capabilities and networks, that cannot be replicated by competitors and will secure a sustainable competitive advantage for the firm.

The purpose of this chapter was to introduce a hypothetical framework of 'new' business models based on previous research. Hereafter comments, views and issues raised in the existing literature about business models and business models innovation were commented upon. The theories grew from a focal and single business view to a more open and network oriented business model view. This was done in order to discover what characterises a new business model.

In the following chapters, cases showing both successful and unsuccessful examples of innovating new business models and how they performed in a Danish context will be presented.

References

- Abell, D.F. (1980), *Defining the Business: The Starting Point of Strategic Planning*, Prentice-Hall, Inc. N.J

- Balachandra, R, (2000), *An Expert System for New Product Development Projects*, *Industrial Management & Data Systems* 100/7, pp.317–324

- Bohn, K & Lindgren, P (2002), *Right Speed in Network Based Product Development and the Relationship to Learning, CIM and CI*, CINet, Helsinki

- Chan, W. K. and R. Mauborgne (2005), Blue *Ocean Strategy, How to Create Uncontested Market Space and Make Competition Irrelevant*, Harvard Business School Press

- Chesbrough, H. (2003), *Open Innovation: The New Imperative for Creating and Profiting from Technology*. Boston: Harvard Business School Press

- Chesbrough, H.W. (2006), *Open Business Models How to Thrive in the New Innovation Landscape*. Boston (Mass.): Harvard Business School Press

- Child, J & Faulkner D. (1998), *Strategies of Cooperation*, Oxford University Press

- Christensen, C. M., (1997), *The Innovators Dilemma: When New Technologies Cause Great Firms to Fail, Harper Business*

- Ulrich, K.T & Eppinger, S.D., (2000), *Product Design and Development*, 2nd edition, Irwin McGraw-Hill.

- Kotler (1994). *Marketing Management Analysis, Planning, Implementation and Control*, Prentice Hall

- Leifer, R December (2002), *Critical Factors Predicting Radial Innovation Success*, Rensselaer Polytechnic Institute, NY

- Linder, J. and Cantrell S. (2000), *Changing Business Models: Surfing the Landscape,* Accenture institute for strategic change

- Lindgren, P. (2003), *Network Based High Speed Product Innovation,*Center of Industrial Production - Buch's Grafiske, Denmark

- Osterwalder, Alexander; Pigneur, Yves; Tucci L. Christopher (2004), *Clarifying Business Models: Origins, Present, and Future of the Concept, Communications of AIS,* Volume 15

- Rosenau, M.D (1993), *Managing the Development of the New Products, ITP*, pp. 39–41

- Tidd, J., J. Bessant, K. Pavitt (2005), *Managing Innovation. Integrating Technological, Market and Organizational Change*, Chicester: John Wiley & Sons

- Vervest P., et al (2005), *Smart Business Networks,* Springer ISBN 3-540-22840-3

- Wind, Y. (1973), *A New Procedure for Concept Evaluation, Journal of Marketing,* Vol. 37, October, pp. 2–11.

Chapter 1.3

ICT – a Key Enabler in Innovating New Global Businesss Models

Peter Lindgren

&

Yariv Taran

Abstract

ICT, information and communication technology appear to be of ever increasing importance to a company's innovation of new global business models. First, companies increasingly use ICT to link into new global networks. Second, their business models, and especially their new business models, increasingly rely on and are enabled by ICT. Third, their innovation is carried out via advanced ICT tools, which means that companies become increasingly dependent on such global ICT networks to extend their business development to the global market.

The ICT network based business model of today includes business models with many types of ICT structures, including both virtual and physical members. Virtual types include e.g. secondlife.com, google.com and facebook.com. Physical models, cooperating from different locations, include: Child 2001, Hollensen 2004, Vervest 2005 and Turban 2008. In a short term perspective, these networks and business models are under a tremendous pressure of cost, speed, performance and change, especially due to the fact that product and business life cycles in the global market are continuously diminishing (Sanchez 2001 and Chesbrough 2007). This has reached a point of no return, where a strong focus on ICT enables companies to cope with this time pressure and the need for agility. In a long term perspective, however, continuous improvement, learning and innovation (Nonaka & Tacheki 1996 and Bohn & Lindgren 2003), in combination with the continuous need to develop new business (Lindgren 2003), increases the importance of ICT to a company's ability to innovate, to stay competitive and to keep the business alive.

Under such conditions, many companies try to solve the innovation and business model challenge by large investments in ICT, especially electronic development tools (e-development). They seem, however, to have done so with varying results (Bohn & Lindgren 2002). This chapter aims to explain some of the importance that ICT has gained in the innovation of new global business models. The ICT component as an increasingly important part of the business model will also be commented upon.

Introduction

The importance of ICT has been growing in the past few decades because of the development of ICT itself. The many rapid changes in markets, technology, networks, and in the competence profile of companies (Nonaka &Takeuchi 1995,

Sanchez 1996, Coldmann & Price 1997, Child and Faulkner 1998, Boer 2001, Lindgren 2003, Christensen 2003; and Turban 2008) have also contributed to an increased demand for a tool that can aid in the speedy integration and facilitation of inter-firm communication. A company that wishes to achieve a competitive advantage on the global market, through innovation, has to develop new business models at an ever increasing speed (Lindgren 2003) and companies focusing on these types of innovation seem to have become increasingly linked and network dependent – either by means of physical, digital or virtual networks (Lindgren 2003) with ICT as an integrator. Two related conclusions can be drawn on the basis of these observations. A company's business model changes rapidly due to the development in ICT - field and business model development becomes more and more facilitated by ICT.

ICT can increase and support the ability to recognise, analyse and develop a new business model and further improve the implementation process of the business plan.

ICT can help business innovation in, at least, four fields. The use of ICT can:

1. Quickly define the task of business model development as being either incremental or radical (W. Chan Kim and Renee Mauborgne 2005)
2. Quickly define the components interacting within the field of where the business model shall operate (market, technology, network and company competences) (Sanchez 2000, Christensen C. 2003, Child & Faulkner 1998, Prahalad & Hamel 1990, Drejer & Riis 2000, Lindgren 2003 and Chesbrough 2007)
3. Quickly define the success criteria of the business model development - short and long term, (Roseneau 1993, Lindgren 2003 and Osterwalder 2005) including percieved cost and perceived values
4. Quickly define the type of business model development best suited to the actual situation such as e.g. stage-gate models, flexible models and rapid prototyping (Clark and Wheelwright 1986, Cooper 1986, Baker & Hart 1999, Verganti 1999, Corso 2002 and Vervest 2005)

ICT is a very important enabler of network based business model development and of the continuous innovation of business models. ICT can be used to achieve a competitive advantage in the field of business model development because it enables high speed development, visualisation of the business model, flexibility, control and cost efficiency.

ICT is, thus, both an enabler of innovation and a component of the business model. This two sided characteristic stresses the importance of ICT in the development of business models.

Theoretical Approach

The e-development enabling role of ICT has been in focus for a long time by virtue of increased and improved technology. Focus has mainly been on software technology (Ronnback 2002) as an enabler to increase speed and diminish the innovation cost at which particular products are developed (Turban 2004). The constant and speedy development of the Internet and related software tools (Chaffey 2006) contribute to making the development of ideas and concepts more precise and rapid, the prototyping easier and time to market more precise.

The e-development enabling aspect of ICT is expected to play a much more significant and important role for companies in future business development than it does today. This will be the case as concerns innovation activities and the establishment of competitive advantages. Especially at the very early stages of business model development, where time or finances do not allow the development of a physical prototype of the business model, an increased use of ICT is obvious (Lindgren 2003, Bennedsen 2007). By means of this it is possible to create an exact virtual copy showing and explaining all of the facilities of the final business model and process. In this way, the business model development process is furthered and it becomes more flexible as well as more cost effective. The supplier and the customer are able to participate in the innovation process and to make important decisions concerning the subsequent course of the business model at a very early stage, and actually, at any stage of the business model process.

The previously encountered physical obstacles to prototyping seem to be overcome, today, by the ICT tools used for business model innovation. Customers, together with suppliers, can continuously innovate the BM and the BM process. The network partners can decide to 'freeze' the whole business model, or a fragmented part of it, in the innovation process, whenever they may so desire. The BM can be transformed into a physical model and introduced to the market when the network partners find 'the right time' (Bohn & Lindgren 2002) to do it. The innovation process can be rewound in order to find out where the innovation process went wrong and why it did so.

The use of ICT is, therefore, related to innovation and to the new business model in various ways. Some companies use e-development tools primarily for their internal product development but this tool can also be put to an advantage in the BM development. Others use e-development in cooperation with sub-suppliers (Lindgren 2003) – but the e-development tool can be used by all network partners joining the model. Other companies use e-development tools externally and in cooperation with their customers only. The e-development tool can be used both within and outside the company, the network and the business model.

For many years, the Danish company Kellpo has developed their capacity for new product development via advanced e-development software tools. Kellpo has brought their e-development to a level where they face economic difficulties when doing business with customers who are unable to develop on an e-development platform together with Kellpo. According to the managing director, Kellpo normally rejects customers who master 'physical' product development only. Kellpo's BM is dependent on ICT and e-development. This could easily prove to be the general practice with innovation of new business models in the future.

Kellpo has developed a competitive advantage by using ICT strategically because customers also come to depend on their e-development platform. Kellpo's customers will, thus, be faced with increased costs on account of switching platforms because of their integration in the e-development platform. Kellpo has positioned itself in a strategic, competitively preferable position. They have created and own the e-development platform used for continuous innovation. This is one possible result of strategically transferring the e-development tool to include the innovation of new business models.

The ability to use the e-development tool and to integrate the network partners in the company's e-development platform is a decisive factor when networks of customers, suppliers and sub-suppliers choose to innovate together. An increasing number of companies will begin to reject suppliers, and even customers, when such partners do not master the e-development tools. The companies can further develop their core competences by using the e-development platform as a strategic, continuous innovation platform for business models.

The research group's proposal is to analyse e-development and continuous innovation related to innovating business models on three dimensions:

1. The internal dimension – e-development taking place inside the company.
2. The collaborative dimension – e-development taking place up and down the stream of the value chain of the companies.

The process dimension – e-development used to control, learn and revise the process of a business model as it is functioning in the marketplace.

The Internal Dimension of E-Development

E-development in the internal dimension concerns 3 levels:

1. The operational level
2. The tactical level
3. The strategic level

The following figure shows an example of the 3 levels of internal e-development and the internal interaction with e-development on each level.

At **the operational level** the e-development challenge is, to a large extent, related to e-development systems, version numbers, language, communication possibilities and barriers. The offer of e-development software is manifold and each software package has its own strengths and weaknesses. Many e-development software providers battle against each other to become the 'e-development software platform' or 'standard' of the future. The use and choice of e-development software is also related to the line of business, product and process task concerned and to the network involved in the business development process.

Figure 1: E-Development Systems and Tools Inside the Company
Source: Lindgren and Bohn 2003

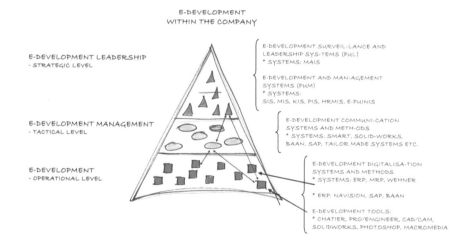

Each e-development system has its strengths and weaknesses that manifest themselves either before, during or after the development process. The challenge for a company is to find its own way through the 'jungle' of e-development tools to improve its abilities, competences and demand for continuous innovation. The market for e-development software today is moving from an innovation phase to a growth phase. In this phase, companies are investing heavily in stable and standardised e-development software and e-development tools.

At **the tactical level**, the e-development challenge relates, first of all, to the way e-development systems enable communication and to the switching of data and knowledge from one system to another. It is very important that the e-development tools can communicate with each other. The 'language' and the 'communication' challenge is by far the biggest barrier to implementation and it is a long term process of internal continuous innovation. Further, it is also vital, and often difficult, to find e-development tools that can communicate outside the company with other network partners.

Many managers at the tactical business development level risk marginalisation because the new e-development tools make it possible to quickly develop close relations with customers and suppliers. Often, there will be no need for tactical management, but only for operational and strategic management. The innovation process can be a method of finding unnecessary members, chains and processes.

At **the strategic level**, the e-development challenge relates to the ability of e-development systems to provide critical data for decision making at the leadership level in order to help strategic managers with the following: recognising the business development task, analysing the field of business development (market, technology, network and companies' competences), defining the success criteria of the business development challenge and in choosing the best suited business development model and process. This area has shown to strongly lack tools, management focus and knowledge (Turban 2008).

The Collaborative Dimension – Up and Downstream Dimension

E-development in the collaborative dimension is related to:

1. The operational collaborative business development level

2. The tactical collaborative business development level
3. The strategic collaborative business development level

The figure below shows the 3 levels of collaborative e-development and interaction at each level.

At **the operational collaborative level,** the e-development challenge concerns e-development systems, version numbers, language, communication possibilities and the barriers faced by each of the participants in a particular network based product development project. The differences between and numbers of e-development software programs are manifold and each network partner has his own opinion regarding the best e-development software tool. Sometimes the solution and tools are chosen on an almost religious perspective, which can be a barrier when starting the process of continuous innovation between network partners. E-development communication between the network partners and the level of communication (simple data exchange or fully integrated systems) are of major concern in most of the NEWGIBM case companies.

The choice of e-development software to be used by collaborative network partners is also strongly related to the line of business, product, process task and the characteristics of the network involved in the business development process. Often, a collaborative network has its own e-development system. It is a challenge for

Figure 2: E-Development Systems and Tools at the Collaborative Strategic Level

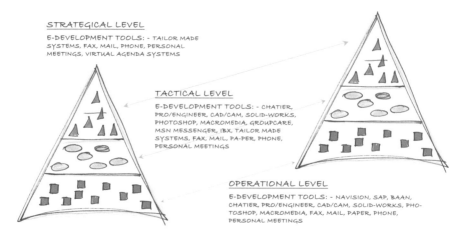

STRATEGICAL LEVEL
E-DEVELOPMENT TOOLS: - TAILOR MADE SYSTEMS, FAX, MAIL, PHONE, PERSONAL MEETINGS, VIRTUAL AGENDA SYSTEMS

TACTICAL LEVEL
E-DEVELOPMENT TOOLS: - CHATIER, PRO/ENGINEER, CAD/CAM, SOLID-WORKS, PHOTOSHOP, MACROMEDIA, GROUPCARE, MSN MESSENGER, IBX, TAILOR MADE SYSTEMS, FAX, MAIL, PA-PER, PHONE, PERSONAL MEETINGS

OPERATIONAL LEVEL
E-DEVELOPMENT TOOLS: - NAVISION, SAP, BAAN, CHATIER, PRO/ENGINEER, CAD/CAM, SOLID-WORKS, PHOTOSHOP, MACROMEDIA, FAX, MAIL, PAPER, PHONE, PERSONAL MEETINGS

the NEWGIBM companies to find a collaborative way through the 'jungle' of e-development tools that improves their abilities, competences and to collaborate and/or improve their continuous innovation together.

Companies are, at present, in an innovative phase in this field, where more and more companies have realised that it is necessary to invest heavily in collaborative e-development. The major challenge in this area is that many collaborative network partners' e-development software cannot communicate with each other or are not flexible enough. Large companies try to solve this problem by forcing their sub-suppliers to invest in specific collaborative e-development tools, such as the IBX– solution. If sub-suppliers reject this 'opportunity', they are strongly at risk of being marginalised and are often phased out as a sub-supplier. However, in a global marked, based on equal networks, the possibility of dictating to network partners which software and 'language to use becomes even more difficult. The e-development software and ICT have to become more intelligent and the companies must define which software and 'language" the chosen network partners have to use in their project handling and communication. The computer company Apple is one of the pioneers in this line of innovation.

Most companies, however, lag behind in the process of continually finding 'translator' e-development software that may make it possible for companies to secure their ability to stay continually attractive as a'hub' network partners can connect to.

It is very difficult and resource demanding for companies to handle this challenge because the 'translator' e-development software is still in the very first phase of the innovative process. It is not, at the moment, possible to predict which software will come to serve as the common standard. There is a huge struggle and battle for IT-providers to find and develop the 'translator' e-development software (EPRI 2007) that will be the standard in the future. Large IT-software providers, such as Microsoft, Oracle and Linux e.g. are investing heavily in this area to become the market leaders or to achieve ownership of the 'translator software' standard. This competition, in combination with the company's choice of collaborative e-development software, seems to influence the degree of innovation and performance of the company and its collaborative network partners in the field of continuous innovation.

At **the tactical level,** the e-development challenge concerns, primarily, the ability of the e-development systems to communicate with and to move data and knowledge from one system to another. Secondly, it is a question of whether business development managers, at a tactical level, perform and improve collaborative management via the e-development tools. Thirdly, it is a question of whether management, at the tactical level, can continuously improve innovation by using e-development on the tactical management level. Presently, e-development, at this level, is only carried out partially. It primarily concerns data communication, the movement and transferral of knowledge and is very poorly developed at the continuous innovation and management levels.

At **the strategic level**, the e-development challenge is related to the abilities of e-development systems to provide leadership with critical data to improve business development managers' decisions. The leadership of the firm continuously needs data and tools to recognise the business development tasks, to rapidly analyse the field of business development (market, technology, network and company competences), to define the success criteria of the business development challenge and to choose the best suited business development model and process.

Companies lack strategical e-development tools at this level. Most communication and collaboration at this level still takes place on a physical basis or via letter, telephone, email and fax. E-development, at this level, is hardly existent and there is a lack of software development in this area to improve future strategical innovation among network partners. The NEWGIBM group is now involved in a development project on this topic together with Stanford Humanities Lab and in the project Smart House.

Summing up on the collaborative e-development area, it can be concluded that the operational collaborative level is developing rather well – but mainly in the product development area, whereas the business development area is advancing at a slower pace. At the tactical and strategical collaborative levels there is a poor or rather non-existent e-development. This seems to be a major barrier to continuous innovation of new business models among network partners.

Development and consensus on technical standards applicable to collaborative e-development progress best upstream of the value chain – between suppliers and the focal company. This means that collaborative e-development primarily takes place in the upstream activities. This seems to be very critical to companies if the drivers for new ideas, new products and the process of continuous innovation, in particular, primarily stem from the customers and the market (Von Hippel

Figure 3: E-Development Up and Down the Stream of
the Collaborative Network and Value Chain

E-DEVELOPMENT IN COLLABORATIVE NETWORKS
– UP AND DOWNSTREAM

2005). When companies are left out of this continuous innovation process, they have a difficult time fitting into the market.

The attractiveness of companies considered from an 'e-development hub possibilities perspective' was another important finding in this field. A company's failing ability to integrate into a new network and to communicate, exchange and share data and information rapidly with network partners is often an obstacle to continuous innovation. Although companies generally try to develop their attractiveness, this is done on a fragmented basis. It seems that companies consider this 'hub' competence critical for their competitiveness judged both from a short and a long-term perspective.

E-development in a Collaborative Process Perspective

E-development has, until today, mostly been a discussion on the digitalisation of the product and the digitalisation of communication, especially within the field of product development. Such digitalisation, however, is of minor importance compared to the digitalisation of the collaborative business development process considered from a competitive and continuous innovation perspective.

The following quotations illustrate this point:

> *'We have realised that one of our competitors has integrated a number of customers into a new e-development platform. The customers were satisfied with this platform and we are now having numerous difficulties in being allowed to develop new products and businesses with these customers,' Jim Larsen, Sales Director from Demex.*

> *'Some of our customers are integrated into a competitor's e-development platform and we struggle to be allowed into their development process,'*
> *L. Mortensen, Managing Director Deluca.*

The e-development in a collaborative process perspective, therefore, concerns 3 levels:

1. The pre-collaborative e-development process level – including all processes of collaborative e-development before the idea or the concept of a new business model is created.
2. The collaborative e-development process level – including all processes of collaborative e-development during the BM and BM process development.
3. The post-collaborative e-development process level – including all processes of collaborative e-development after the BM and BM process has been introduced to the market.

The Pre-collaborative E-Development Process

The pre-collaborative e-development process level includes a company's ability to handle the e-development process on an operational, a tactical and a strategical level before the idea or the concept of a new product is created. Companies can gain access to innovative, new ideas and concepts at the very beginning of the BM development, if they integrate advanced e-development tools at this stage. Companies can gain valuable time if their e-development tools are able to find such ideas and concepts that can bring new BMs closer to wining 'the race of first mover advantage'.

Figure 4: The Three Levels of Collaborative
E-Development Process Development

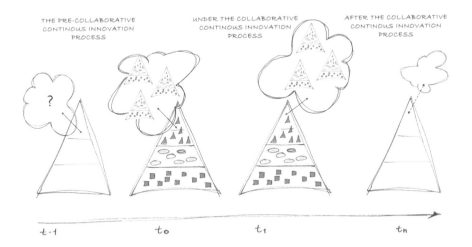

Under the Collaborative E-Development Process

Under the collaborative e-development process, the focus is on a company's ability to handle the e-development process during the BM and the BM process development.

It is important for companies to use ICT, both at an operational, tactical and strategical level to bring the company in a position of ownership of the innovation process in relation to both customer and suppliers. Companies that own the innovation process also own and control the innovation of the customers and the suppliers. This will prevent competitors from taking over customers and suppliers simply because of the switching over costs being too high because they depend on the continuous innovation e-development platform.

The Post-collaborative E-Development Process

This stage includes the whole process of collaborative e-development after the BM and the BM process have been introduced to the market. The post-collaborative development process is the most important part of the process related to continuous innovation and learning success criteria. Companies that control this phase can strategically, tactically and operationally spot the right time, the right cost and the right performance of developing new BMs and BM processes to customers and suppliers. Companies can develop a special ability to wait for the new ideas and concepts of the customers, the competitors and the suppliers in the post-collaborative e-development process. Via e-development tools, it will be possible for them, together with their customer and suppliers, to find the optimal moment for entering the market with a new business model.

Summing up the process of e-development, it appears that:

- It is important for companies to carry out and develop e-development on all levels before, during and after the BM innovation process in order to control and to ensure involvement in the continuous innovation process of the BM.
- It is important for companies to use and to develop e-development. However, it is more important to focus on use and development of collaborative process e-development because this will enable the company to own the process and, in consequence, to own the BM development process of their customers and suppliers. From a competitive advantage perspective, the NEWGIBM case research showed that companies relying on ICT, in the described way, seem to obtain major competitive advantages and have a higher performance in the field of continuous innovation.

Discussion on Future Challenges for E-Development Concerning Continuous Innovation of Business Models

Much has been written about the e-development enabler The focus has mostly been on the e-development enabler as a 'smart' software tool to help companies reduce time to market and minimise cost of product development. Reduced time to market, increased market preference and frequent technical changes on all fronts have, however, placed managers of industrial firms in a difficult situation because the most important competitive advantage of the e-development enabler is not related to reduced time and cost – but to value creation and strategic leadership of the business innovation process. Reduced time and cost is no longer enough to optimise the collaborative development process. The real advantage of e-development, seen in a continuous innovation perspective, lies in the ability and potential for companies to strategically own the collaborative development process. Today, focus is changing in many companies to the abovementioned and to a higher focus on performance (value), continuous improvement, learning and continuous innovation. Innovation management is, therefore, forced to create and integrate the company into a continuous innovation process of new and updated e-development processes with customers and suppliers in order to stay attractive, to stay competitive and to be involved in continuous innovation. As e-development resources are limited and the uncertainty of technical standards is considerable (Cheng & Van de Ven 1996), (Chesbrough 2007), management on tactical and strategic e-development levels need to deal with the uncertainty, risk and strong demand of continuous innovation in the field of product development. Therefore, innovation leadership and management in networks and, consequently, outside the boundaries of organisations offer new e-development challenges. These challenges are complicated as organisations may differ over the time of a business development project process related to continuous innovation. Commitments e.g. to the process may influence collaboration, as one single partner will not have authority over another partner. This underlines the importance of the need to take into consideration the role of ICT and e-development tools, in both the BM and the BM innovation process. This should be done both at a leadership and at a management level in the collaborative new business development.

Conclusion

A company's ability to gain competitive advantages in respect to the continuous innovation of their BM and innovation of new BMs is related to the ownership or the integration of the e-development platform. E-development in collaborative business development projects can be related to the following elements: internal e-development, external collaborative development and the collaborative e-development process. There is a strong relationship at each of the three levels between the use of e-development in collaborative network business development and the performance of the new business model.

The use of e-development as an enabler in the process of both innovation and business development is of major importance for companies today. E-development can bring companies in a strategic position of owning the customers' innovation process.

The discussion shows the importance and also the implications regarding e-development in collaborative BM development. The choice of e-development tools, the choice of ICT tools and the choice of e-development network structure and platform have a significant influence on the innovativeness and performance of business development and continuous innovation of new business models.

References

- Baker, M. & Hart, S., (1999), *'Product Strategy and Management'*, Prentice Hall, Harlow, U.K

- Balachandra, R., (2000), *'An Expert System for New Product Development Projects'*, Industrial Management & Data Systems 100/7, pp.317–324

- Bessant, J., 81999) *'Challenges in Innovation Management'*, Centre for Research in Innovation Management, University of Brighton

- Boer, H., (2001), *'Operational Effectiveness, Strategic Flexibility or Both? – A Challenging Dilemma'*

- Bohn, K. & Lindgren, P., (2002), *'Right Speed in Network Based Product Development and the Relationship to Learning'*, CIM and CI', CINet, Helsinki

- Caffyn S.J., (1998), *'The Scope for the Application of Continuous Improvement to the Process of New Product Development'*,University of Brighton

- Child, J. & Faulkner D.,(1998), *'Strategies of Co-operation – Managing Alliances, Networks, and Joint Ventures'*, Oxford University Press, Oxford

- Christensen, C (2003), *'Disruptive Technology'*

- Corso, M. et al., (2000), *'Knowledge Management in Product Innovation: An interpretative Review'*, International Journal of Management review Vol. 3, N. 4,pp. 341–352

- Corso, M., Martini, A., Pellegrini, L. & Paolucci, E., (2001), *'Knowledge Management in SMEs – Does Internet Make a Difference?'* Italy

- Day, G.S., (1990),, *'Market Driven Strategy – Processes for Creating Value'*, The Free Press, New York

- Drejer, A. & Riis, J.O., (2000), *'Competence Strategy'*, Børsens Forlag, ISBN 87-7553-740-0.

- Gieskes, J., (2001), *'Learning in Product Innovation Processes – Managerial Action on Improving Learning Behaviour'*, ISBN: 90-365-1651-x

- Goldman, Nagel & Price, (1998), *'Agile Competitors and Virtual Organisations'*, Van Nostrand Reinhold, New York

- Håkansson, H., (1987), *'Product development in Networks" In: Håkansson,H* (ed.) Industrial Technological Development. A network Approach, Croom Helm, London, pp 84–127

- Hørlück, J., Kræmmergaard, P., Nielsen, J.S. & Lindgren, P., (2002), *'Interorganizational Project Management'*, CINet, Helsinki

- Hørlück, J., Kræmmergaard, P., Rask, M., Rose, J. & Lindgren, (2001), *'Organizing for Networked Information Technologies – Cases in Process Integration and Transformation',* Aalborg University Press and PITNIT, Denmark

- Laage-Hellmann, J., (1989), *'Technolgical Development in Industrial Networks'*, Uppsala University 1989

- Leifer, R., December, (2002), *'Critical Factors PredICTing Radial Innovation Success'*, Rensselaer Polytechnic Institute, NY.

- Lindgren, P., (2003), *'Networkbased High Speed Product Development Models and Processes,* Aalborg University

- MacCormack A. & Verganti R., (2002), *'Managing Uncertainty in Software Development: How to Match Process and Context'*

- MacCormack A., Verganti, R. and Iansiti, M., January (2001), *'Developing Products on 'Internet time': The Anatomy of a Flexible Development Process'*, Management Science/Vol. 47, No, 1.

- MacCormack, A. & Iansiti, M., (1997), *'Product Development Flexibility'*, 4th International Product Development Management Conference, EIASM, Stockholm, Sweden

- Nonaka, I., & Takeuchi, H., (1995), 'The Knowledge-Creating Company', New York: Oxford University Press, pp. 93–138

- Prahalad, C. K. & Hamel, G., (1990), *'The Core Competence of the Corporation'*, Harvard Business Review, May-June, Vol. 68

- Riis, J.O. & Johansen, J., (2001), *'Developing a Manufacturing Vision'*, International Working Conference on Strategic Manufacturing, Denmark

- Rosenau, M.D., (1993), *'Managing the Development of the New Products'*, ITP, pp. 39–41

- Ronnback, Anna Ohrwall., (2002), *'Interorganisational IT Support for Collaborative Product development'*, Unitryk, Linkøpings Universitet ISBN. 91-7373-323-7

- Sanchez, R., (1996), *'Strategic Product Creation: Managing New Interactions of Technology, Markets and Organizations'*, European Management Journal Vol. 14. No 2, pp 121–138.

- Sanchez, R., (2000), *'Product, Process, and Knowledge Architectures in Organizational Competence'*, Research Working Paper, Oxford University Press, 2000-11.

- Sanchez, R., (2001), *'Modularity, Strategic Flexibility, and Knowledge Management'*, Oxford University Press.

- Saren, M.A., (1984), *'A Classification and Review of Models of the Intra-Firm Innovation Process'*, R&D Management, 14 (1): 11–24

- Smeds, R., Olivari, P. & Corso, M., (2001), *'Continuous Learning in Global Product Development: A Cross-Cultural Comparison'*, Finland.

- Turban, E et al,, (2000), *'Electronic Commerce a Managerial Perspetive'*, ISBN 0-13-975285-4

- Tseng, M.M., Jiao, J., & Su, C-j., (1998), *'Virtual Prototyping for Customized Product Development'*, Integrated Manufacturing Systems, 9/6, pp. 334–343.

- Ulrich, K.T & Eppinger, S.D., (2000), *'Product Design and Development'*, 2nd Edition, Irwin McGraw-Hill.

- Verganti, R., MacCormack, A. and Iansiti, M., (1998), *'Rapid Learning and Adaptation in Product Development: An Empirical study on the Internet Software Industry'*, EIASM 5th International Product Development Management Conference, Como, Italy (25–26 May 1998).

- Wheelwright, S.C. & Clark, K.B., (1992), *'Revolutionizing Product Development. Quantum Leaps in Speed, Efficiency, and Quality'*, Free Press, New York, NY.

Chapter 14

The NEWGIBM Research Methodology

Jacob Høj Jørgensen

René Chester Goduscheit

&

Carsten Bergenholtz

Introduction

The aim of this chapter will be to present how the NEWGIBM research frame-work was built – or more precisely, a conceptual framework – that will allow an in depth analysis of a number of case studies of firms that are innovative in a network. The methodology in the development of this framework is to a large extent, inspired by the seminal work of (Eisenhardt 1989) and, later on, by (Yin 1993) and e.g. (Halinen & Törnroos 2005). It is the intention of this chapter to focus primarily on the methodological aspects of the NEWGIBM research and not on the results.

The research project NEWGIBM (New Global ICT based Business Models) cooperated closely with a group of firms during 2005 and 2006. The focus of the project has been development of new business models (and innovation) in close cooperation with multiple partners. These partners have been customers, suppliers, R&D partners, and others. The methodological problem is thus, how to come from one in-depth case study – or a number of case studies – to a more formalised theory or model on how firms can develop new projects and be in-novative in a network.

The chapter is structured so that it starts with a short presentation of the case study methodology and the research setting with three fundamental research questions regarding the methodology addressed. The remainder of the chapter presents a distinction between out-come driven and event-driven research and makes a case for the latter as the most relevant.

One of the case studies and the focal firm will be presented later in this chapter together with a presentation of the focal firm followed by a detailed descrip-tion of how the research process has been organised. This is followed by a discussion of how to analyse a single case study. In the remaining part of the chapter a more general discussion of how to develop a valid and operational theory from a single case study is presented with some examples from the actual case study.

The Case Study Method

The case study method as in e.g. (Eisenhardt 1989; Eisenhardt & Graebner 2007; Halinen & Törnroos 2005; Siggelkow 2007) or in the seminal publications of Yin, as e.g. (Yin 1993) is well known. A case study is typically seen as a research design that focuses on a single setting or unit that is spatially and temporally bound. In each case, several sub-cases can be embedded (Yin 2003). Yin (1993) defines a case study as an empirical enquiry into a phenomenon in its real-life context where the boundaries between the phenomenon and the context are not clearly evident. Eisenhardt (1989) stresses the potential of a case study to capture and present the dynamics of the phenomenon studied in the actual context. A case study gives depth and comprehensiveness to the research instead of partial explanations, see e.g. (Easton & Håkansson 1996).

The case study method is an effective method for the study of changes due to the combination of contextual factors and process elements in the same real-life situation. The case study method is, therefore, often used in the analysis of innovation processes as in (Bessant & Caffyn 1997; Bonaccorsi & Lipparini 1994; Chesbrough 2003; Christensen, Anthony, & Roth 2004; Haryson 2000). A rich description and analysis of a single case can be a valuable tool in generating a valid theory, see e.g. (Eisenhardt 1989; Glaser 1992; Glaser & Strauss 1967; Miles & Huberman 1994). Even if a case study is a valuable method for generating theory, it raises several problems to be dealt with. The field of study is complex and gets even more complicated by the temporal considerations that are necessary to grasp the dynamic problems of real life.

The Research Setting

Halinen & Törnroos (2005) studied the development of networks by distinguishing between individual and context and by looking at the history and the future of the network when studying it from its actual position. Inspired by this methodology, while not studying the history of the network, the authors have been able to follow the networks from their first steps through several meetings, crisis, etc. up to present time. The initial model of the research framework can thus be seen as in figure 1.

In accordance with the discussion above, a number of problems arises which have been necessary to consider during the process:

- The complexity problem – how does one separate the focal activities from the context – when is there talk of an innovation activity?
- The time problem – how does one cope with the dynamics of a network that is continuously evolving and thus, constantly changing its nature and its relations to the context, which is developing too?

Figure 1: First Research Model Inspired by (Halinen & Törnroos 2005)

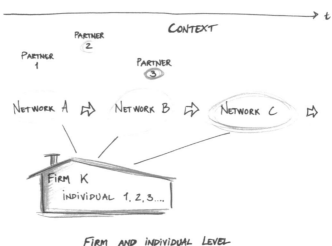

* The validity problem – how is it possible to develop a valid theory and an operational model through a large scale case study like this?

The 'time' problem will be dealt with and then a research design, which can be labeled 'event-driven' will be proposed. Next the complexity problem will be discussed, which can be seen as linked to the boundary problem. At last the validity problem will be addressed.

Out-Come-Driven or Event-Driven Research

Outcome driven research or event-driven research are two classical research strategies in the social sciences, especially when dealing with case studies, see e.g. (Poole et al. 2000; Teece, Pisano, & Shuen 1997; Van De Ven & Engleman 2004). Explanations that are outcome driven are typically built from observed outcomes to prior events that are seen as significant. An innovation in a firm is reconstructed as a process of significant events leading to this innovation. Typically, a large number of concepts or variables that could influence e.g. an innovation are examined and evaluated; see e.g. (Van De Ven & Engleman 2004) and (Aldrich 2001). In this way, independent variables can be evaluated statistically as to the extent in which they can explain variations in the outcome, e.g. the degree of success of an innovation. Outcome driven research is useful if there is a well-defined set of variables – a few dependent variables and a large, but fixed, number of independent variables.

The alternative is to use an event-driven approach. In this approach, the research strategy is to follow the process from one event to another and to the final outcome. This means that the research will often be reconstructed as a narrative. In this way, the researcher tries to explain both how and why things happened in the way they did. This way of doing research is typically associated with the analysis of changes in an organisation; see e.g. (Abbott 1990; Gersick 1994;Poole, Dooley, Van de Ven, & Holmes 2000; Van De Ven & Engleman 2004). The problem will typically be how to define the characteristics of an important event in the process towards the outcome. Outcomes may have sub-outcomes and, consequently, later outcomes are themselves events, with subsequent consequences and so on. Another problem is that this kind of research demands access to the organisation over a long period of time.

The empirical research presented in the NEWGIBM project is clearly event-driven as the result – the outcome – is not yet known. Instead the process was followed closely, e.g. through observation by taking part in all meetings within the organisation and in those among the organisation and the external partners.

Presentation of One of the Case Studies: KMD
The case of KMD will be used to illustrate how the research process has been developed as well as the methodology that has been used. The firm will be referred to as K. The intention is to give a short presentation of the firm, its new business model and the development of this model during 2006.

K is a Danish IT company, which provides solutions for public and private companies. The solutions range from operating systems for public organisations to ERP solutions for large private companies. For the last few years, part of the strategy of the company has been to increase the customer scope to include the energy supply sector as well. As of ultimo 2005, the energy division has an annual turnover of 13 m€ and 65-70 employees. As a point of departure, K was operating in the B2B market. Since 2001, however, the company has been a part of a consortium providing an electronic solution for Danish citizens. In 2006, the number of users for this solution was approximately 400,000.

The business model of K was based on the vision of merging its resources within the energy sector and its proximity to a large number of private households through the use of an internet based system called e-Box. The use of e-Box would ensure the initial penetration into the consumer market. In this way, the business model for K could include achieving ownership of the energy information of the consumer market.

The network started in December 2005 at a meeting between two representatives of K and two academic researchers. The initial thought was to provide the consumer market with solutions, that could help Danish consumers to survey and control their energy consumption. It was clear for K that, in order to reach a wider range of private energy end-users, it had to collaborate with other organisations. The K representatives agreed on trying to set-up a network of organisations with a view to develop ICT based energy solutions. Through the next few months, after the initial meeting, K contacted (by mail and email) and visited potential partners for the network.

In March 2006, the entire network met for the first time. The participating organisations were K, two utility providers, a software developer, a telecommunications company and a provider of specific devices for energy consumption. The meeting was setup as a brainstorming seminar, in which all the participants could give their input about potential intelligent home solutions.

The next network meeting, in August 2006, took place at the conference facilities of an external provider within process consultancy and with special expertise on innovation. The meeting was arranged as a combination of plenary discussions and workshops between the participants. The participating organisations were the same as at the first meeting in March, but, as well as, an ICT consultant, an insurance company and a provider of sockets.

In October 2006, K and four utility providers (out of which one organisation participated at both of the meetings mentioned above) joined forces in developing the energy solution for the customers of the utility provider. The solution was to be based on an internet-mediated IT portal, which K currently employed for the end-users within other areas than the utility sector.

Analyzing a Single Case Study

The purpose of the case study was to obtain a rich and deep foundation of data describing the dynamics and processes in the development of K's innovation project. A prerequisite for fulfilling this objective was getting as close to the data as possible. In this case study, it meant attending as many meetings as possible, both network meetings and internally at K.

Figure 2 describes the process of the case study of K. The initial idea was K's, prior to any researcher involvement. The figure illustrates the research process applied

Figure 2: Model of the Case Study Process

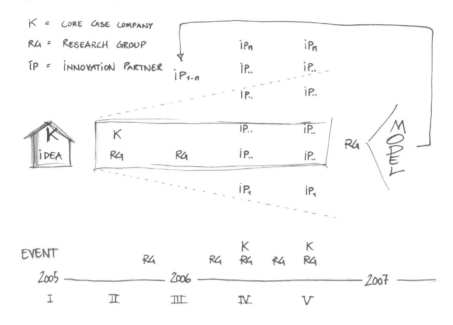

when gathering data, validating data and building a new theoretical model on network-based innovation processes.

The model illustrates who participated in which events as well as the ongoing development of the mental framework (the dotted lines), related to the meetings and seminars (events). The mental framework should be understood as the perspective of the Research Group ("RG"). It is, thus, the foundation for all the work that is done, but a foundation that is everchanging. The RG began the case study with an idea of the kind of innovation K was about to carry out, and what the RG should observe and analyse. This was more a conceptual starting point and involved preliminary models, rather than an actual theory. Different aspects that initially were not considered significant revealed themselves as highly significant. Different levels in K's way of leading the innovation process, for instance, were discovered, and the subtle interplay between the Innovation Partners (IP) appeared to have considerable influence on the process. The expanding mental framework is illustrated with the two dotted lines as in contradiction to the initial mental framework of the researchers illustrated with the rectangular box.

The Complexity Problem

Naturally, the quantity of data increased during the process. But, as the above examples illustrate, the expanding framework also entails that data, that were not considered significant during event 1, might suddenly seem significant after event 3 (this aspect will be elaborated in following sections). Each event challenged the present mental framework and provided a new understanding of prior events. Thus, the expansion of relevant data was both due to the natural accumulation and the developing framework.

Current events will affect the perspective on previous events. If one asks himself or the participants a question relating to event 1 at event 2, he might get a different answer from what he would get if he waited until after event 3. He might even want to frame the question differently. The time perspective has thus revealed itself to be decisive, not only for the researchers, but also for the involved participants in the network.

At least two researchers were present at all interactions, and the framework was discussed formally at debriefings following all events, and informally during the meetings and in between meetings. Minutes were taken at every debriefing and these minutes illustrated, that the mental framework evolved significantly.

When the case study finished, the RG had developed a preliminary model for network-based innovation processes. This model can be partially tested by interviewing key personnel involved in the network. The purpose of interviews is to verify that the RG, K and IP agree concerning critical observations and interpretations. The RG, however, has to consider the time perspective and the fact that respondents might answer differently now in comparison to responses made during the actual innovation process. These methodological points about the development of framework, the accumulation of data and the time perspectives underline the complexity of the case study and the methodological considerations involved. Further content related complexity issues are the intricate processes, key concepts, the technologies and business models involved, strategic considerations and interorganizational relations, just to name a few. Taking into account that the process is event-driven, it would have been difficult to choose an applicable model before the case study was initiated. The purpose of the initial framework is, to be precise, to reduce this complexity and emphasise the aspects that are relevant for the group's research. When such a reduction of complexity is performed, it is critical to be aware of alternative explanations. The RG had to choose between alternative explanations. These interpretational choices are vital and affect future interpretations; in Figure 3 different dotted lines can thus be outlined. The perception of reality is affected since researchers with different mental frameworks will ask different questions.

A Valid and Operational Theory

The vital criterion concerning the interpretational choices is credibility. An open and reflective attitude concerning the use of data and theory is required in order to be able to explain what each statement is based on. It is essential that the RG is aware of the fact that very different perceptions and interpretations are possible. Roughly speaking, it must be made clear that the concluding statements are based on both the types of data (which observations, respondents remarks etc.) as well as the kind of model or theory. This methodological approach provides the most credible results. It will be possible to explain why different kinds of interpretational choices have been made and to identify the facts, perceptions, observations and theoretical insights, which these choices are based upon.

The traditional research dichotomy between an inductive and a deductive research strategy (which is often seen in case study research) isof no use in this research setting. Instead of seeing existing theories and models as something to be tested, they must be seen as part of the context for the case study. The case study takes place in a cultural, social, and economic setting, which has to be part of the analysis and interpretation. In the same vein, the theoretical settings must be seen as part of the context for the case study. Inductive and deductive research strategies must be part of the same process to ensure the validity of the study.

In this way, validity means to ask questions e.g. regarding the connection between the theoretical frame and the questions asked in the interviews. Central conclusions that are extrapolated from the report must be shown to be based on the data and in the theoretical framework to be valid. A reader must be able to judge – and validate – a study on its own premises by judging the methodological procedure (How the case study has been carried out) and the theoretical frame or context.

Conclusion

The problem addressed in this chapter has been the question of how to develop models and theories that are valid and operational from a single case study of innovation in networks. It is shown in the chapter that the interpretation of the events in the case study, to a large extent, depends on the theoretical definition of focal concepts. One of the main conclusions is, thus, that a pure inductive method – like in classical Grounded Theory – is not a reasonable research strategy. Instead, the research strategy is a mixture of 'grounded' interpretations of the events, such as meetings in the network, and more deductive analysises of the development of the case.

The reason behind the case study method is to capture and present a real life phenomenon in the actual context encompassing as many of the dynamics as possible. Depth and comprehensiveness are thus the key elements in a case study. The case study method raises several problems dealt with in this specific research setting of innovation in networks. These problems can be summarised as: The complexity problem, the time problem and the validity problem.

Regarding the time problem – how to analyse a dynamic case – the authors pointed to the difference between out-come driven and event-driven research. The research setting of the case study is clearly event-driven as the process is followed from one event to another and to the final outcome building a narrative of the development of the case. The problem with this research setting is, however, that it can be difficult to define exactly what the constitutive factors of an important event are – or how to distinguish between event and context. An attempt was made to solve this problem by going back to the research field and re-interviewing the central actors, both from the company in focus and from the network partners. In this way both the data from the case and the interpretation of the research group have been validated. This methodology solves some of the validity problems of the case study method and, to some extent, the complexity problem.

In trying to understand the methodological process of this case study – and the other case studies in the NEWGIBM project – the mental framework of the researchers in the research group is extremely important. Understanding the processes behind the development of this framework can be seen as a sort of 'meta' reasoning regarding the researchers' own way of seeing the case and interpretation of events in the case. New concepts like 'innovation leadership' had to be elaborated on during the process because of actions taken by the participants in the innovation process and because of the development of the mental framework of the researchers.

The complexity of the case study, of course, dramatically increases during the above mentioned process and some kind of data reduction is often necessary. In the aforementionned research setting, this reduction is brought about by the research group (two senior researchers, two PhD students and two research assistants) during its meetings and not by the individual researcher. The reflectivity of the researchers has been a key issue during the process and is still. When the final model, built on this and other case studies, is presented, three things should be clear to the reader how the model is grounded in the data from the case studies, how the theoretical frame has been developed, and how the methodological procedure has been applied. In this way, the model can be called valid and operational.

References

- Abbott, A. 1990, *"Conceptions of Time and Events in Social Science Methods: Causal and Narrative Approaches"*, Historical Methods, vol. 23, no. 4, pp. 140–150.

- Aldrich, H. 2001, "Who Wants to Be an Evolutionary Theorist?", *Journal of Management Inquiry*, vol. 10, no. 2, pp. 115–127.

- Bessant, J. & Caffyn, S. 1997, "High-involvement Innovation Through Continuous Improvement", *International Journal of Technology Management*, vol. 14, no. 1, pp. 7–28.

- Bonaccorsi, A. & Lipparini, A. 1994, "Strategic Partnerships in New Product Development: An Italian Case Study", *Journal of Product Innovation Management*, vol. 11, no. 2, pp. 134–145.

- Chesbrough, H. W. 2003, *"Open innovation: The New Imperative for Creating and Profiting from Technology"*, Harvard Business school press, Boston.

- Christensen, C. M., Anthony, S. D., & Roth, E. A. 2004, *"Seeing What's Next: Using the Theories of Innovation to Predict Industry Change"* Harvard Business School Press.

- Easton, G. & Håkansson, H. 1996, "Markets as Networks: Editorial Introduction", *International Journal of Research in Marketing*, vol. 13, pp. 407–413.

- Eisenhardt, K. M. 1989, "Building Theories from Case Study Research", *Academy of Management Review*, vol. 14, no. 4, pp. 532–550.

- Eisenhardt, K. M. & Graebner, M. E. 2007, "Theory Building from Cases: Opportunities and Challenges", *Academy of Management Journal*, vol. 50, no. 1, pp. 25–32.

- Gersick, C. J. G. 1994, "Pacing Strategic Change: The Case of a New Venture", *The Academy of Management Journal*, vol. 37, no. 1, pp. 9–45.

- Glaser, B. G. 1992, *"Basics of Grounded Theory Analysis"*, Sociology Press, Mill Valley, CA.

- Glaser, B. G. & Strauss, A. 1967, *"The Discovery of Grounded Theory"*, Aldine, Chicago.

- Halinen, A. & Törnroos, J. Å. 2005, "Using Case Methods in the Study of Contemporary Business Networks", *Journal of Business Research*, vol. 58, no. 9, pp. 1285–1297.

- Haryson, S. 2000, *Managing Know-who Based Companies: A Multinetworked Approach to Knowledge and Innovation Management"*, Edward Elgar, Cheltenham.

- Miles, M. B. & Huberman, A. M. 1994, *"Qualitative Data Analysis"*, 2 edn, Sage, Thousand Oaks.

- Poole, M. S., Dooley, K., Van de Ven, A. H., & Holmes, M. E. 2000, *"Organizational Change and Innovation Processes Theory and Methods for Research* Oxford Univ Press.

- Siggelkow, N. 2007, "PERSUASION WITH CASE STUDIES", *Academy of Management Journal*, vol. 50, no. 1, pp. 20–24.

- Teece, D. J., Pisano, G., & Shuen, A. 1997, "Dynamic Capabilities and Strategic Management", *Strategic Management Journal*, vol. 18, no. 7, pp. 509–533.

- Van De Ven, A. H. & Engleman, R. M. 2004, "Event- and Outcome-driven Explanations of entrepreneurship", *Journal of Business Venturing*, vol. 19, no. 3, pp. 343–358.

- Yin, R. K. 1993, *Applications of Case Study Research* Sage.

Chapter 1.5

The Analytical Model for NEWGIBM

Peter Lindgren

Yariv Taran

&

Anders Mcllquham-Schmidt

Abstract

The chapter describes how the explorative analysis component of the analytical framework for the NEWGIBM project was set up. The analytical framework presented here is based both on previous findings and results as well as on more recent findings. The purpose of the framework is to attain a visual and holistic recognition of the new global (network-based) business model function, and to find the drivers that cause the network partners to innovate such a business model in the first place. As stated earlier, an understanding of how such drivers work is critical. They are often very complex to codify, because they can also manifest themselves in a non-physical form and value – and in some cases they are related to digital and virtual values.

Keywords: Business model innovation, new global business models, network based business model, framework for NEWGIBM.

Introduction

Prior to this research, it was clear to the authors that this study, as well as its analytical framework, would be explorative and a challenge to pursue. The reason for this was mainly the fact that the research dealt with a network-based BM formation, which was characterised firstly by its various network formation ties and structures, and secondly by values and perceived values - which inevitably also extended the views on customers' identity and BM's definition.

In spite of the progress made in business model research thus far, the focus has mostly been limited to that of individual firms. Today, however, it is commonly understood that we live in a network economy. Consequently, the business model concept where the BM serves merely as '[a] blueprint of how a single company does business', as suggested previously by Osterwalder et al. (2004) is being challenged. Instead, it is increasingly perceived as a blueprint that explains how network partners do business together, in a platform, where each partner can strengthen its own weaknesses by exploiting other network partners' competencies and skills. The partner can also modify its own competencies by fine-tuning with the core competencies of other network partners so as to achieve synergetic, network-level benefits.

With attempting to tackle this issue, the NEWGIBM group began to develop a new framework for analysing new global business models, a new research strategy and a new research methodology toolbox.

Background of the Analytical Model

In the previous chapters, it has been clarified when a BM can be defined as 'new'. The research, however, appears to show that the success criteria of the New BM structure are strongly related to another important key factor, which has not been taken into consideration until now – the 'Network-Based' configuration of firms.

Yet, this did not provide an analytical framework that could be used to analyse the NEWGIBM - cases. As mentioned earlier, the biggest challenge in the research project was to close the gap between the intellectual environment, where authors with academic backgrounds generally have no trouble spinning theories on various business models' contribution to organisations, and the business leaders, who are more interested in finding the proof that those theories actually work in practice.

For that reason, it was decided that the NEWGIBM group should first develop a new analytical framework in order to establish an understanding of the BM concept, and next methods for practitioners, case companies, and network partners to increase their innovation potential and develop more successful new global business models should be defined.

As part of the authors' preliminary research, the following important trends and characteristics in the literature on global business models innovation were found.

Table 1: Economic Implications of a BM's Innovation

Context for Innovation	Until Today	Trends for the Future
Market	National. Stable. Common. Mainly physical.	Global. Fragmented, Dynamic, Customised. New markets (Blue Ocean). More digital and virtual.
Technology	Single technology. Expensive. Data power low. Stable.	Mix of technology or multi-technology. Cheap. Data power over capacity. Unstable - Rapid new technology changes.

Context for Innovation	Until Today	Trends for the Future
Network	Closed networks. Stable networks.	Open networks, Dynamic networks, Virtual networks, Global networks.
Companies' Competences	Stable competences developed inside the company or in a narrow network.	Dynamic – flexible competences. Competences continuously developed under pressure. Competences developed with many network partners - sharing core competences and skills in the innovation process (to reduce the risk of disruptive technological changes within the industry).
Product	Mostly physical products To some extent immaterial products. Stable product – long life cycle. Limited distribution and marketing channels.	A mixture of physical, immaterial, digital and virtual products. Continuous development of product - short life cycles. Many distribution and marketing channels.
Product Innovation Model	Stable models. Slow, linear innovation process.	Many product innovation models. Lean product innovation process. Flexible models, Dynamic models. Rapid prototyping models (learning by fast experimentation).
Success Criteria	Individual success, innovation speed, time to market, cost and performance, local market Emphasis on short term success criteria. More emphasis on CI and managing tangible assets efficiently.	Network-based success, innovation speed, time to market, cost and performance, global markets. Emphasis on short and long term success criteria. More emphasis on radical innovation. More emphasis on managing intangible assets efficiently.

As can be seen from the table above, the rapid changes in the economy are changing the rules of the game[1] when dealing with BMs. Such changes are creating an increase in competition and companies must learn to adjust to those changes – and quickly. The new knowledge based economy is very chaotic and fast moving

---

1   Lundvall, B.A. (2002)

and the future trajectories are more difficult for the companies to predict. It is a lean based and, therefore, also a customer driven economy. Globalization and competitive advantage are being strongly related to differentiation[2] and continuous innovation within networks.

Consequently, competition today is becoming increasingly difficult for companies trying to face it alone. Therefore, in addition to developing their core competence base individually, enterprises have to become more open toward network-based innovation, particularly when the core knowledge and competencies needed to improve their performance are not available in-house. Main potential sources of external competencies are customers and suppliers, consultants and even competitors. Collaboration helps them find and open new markets, and obtain a better understanding of what is expected from them (Barclays 2002). Consequently, many companies are finding themselves increasingly tied to other firms. The challenge each of them faces is to adjust their own business model to meet their partners' core competencies so as to create a new and joint platform for collaboration and innovation.

In view of this problem, the case studies conducted by the authors suggest that it is useful, not only to study the business model concept at the level of individual firms, but also at that of networks. This, however, calls for some adjustments to the business model building blocks suggested previously by Osterwalder et al. (2004). In Table 2 below, these network-level building blocks are summarised.

Table 2: Building Blocks of a Network-Based Business Model

Pillar	Network-Level Business Model Building Blocks	Description
Product (physical, digital, virtual)	Value proposition	Gives an overall view of each network partner's bundle of products and services
Process (physical, digital, virtual)	Process structure	Gives an overall view of each network partner's bundle of processes

---

2   Bohn & Lindgren 2003; 'Vision of the future' (1995)

Pillar	Network-Level Business Model Building Blocks	Description
Customer Interface (physical, digital, virtual)	Target customer	Describes the customer segments that each network partner wants to offer value
	Distribution channel	Describes the various means used by each network partner to get in touch with its customers
	Relationship	Explains the kind of links established between each network partner and its different customer segments
Infrastructure Management	Value Configuration	Describes the arrangement of activities and resources involved in developing the new business model by each network partner
	Core Competence	Outlines the necessary competencies for each network partner in order to execute its business model
	Partner Network	Portrays the network of cooperative agreements (network construction) with other network partners necessary to efficiently offer and commercialise value.
Financial and Non-financial Aspects	Cost Structure	Sums up the monetary and non-monetary consequences of the means employed in the business model
	Revenue Model	Describes the way each network partner makes or intends to make money or create other non-financial values

In relation to the intra-firm version, the process dimension and the non-financial aspects were added and some of the building blocks were specified in more detail. Otherwise, the elements are essentially the same. The main difference concerns the unit of analysis – the network instead of the individual firm. Furthermore, the earlier framework did not consider any perceived value/cost or the perspective of joint innovation. Neither did it specify when a business model could be defined as new from a network-based point of view.

In view of that, as will be seen later in the book, the case studies suggested the following differences between firm-level and network-level business models:

- A network-based business model is a powerful platform for individual companies to innovate from, particularly when they recognise that the core knowledge and/or competencies needed are not available in-house. This point is applicable to all phases of the innovation process from idea generation to application.
- The level of innovation is potentially high within a network-based business model, due to the large mixture of competences and ideas that become available to the partners.
- The 'time to market' of a network-based innovation can be reduced significantly compared to single-firm innovation, either because of an immediate exploitation of existing technologies, of faster market introduction, or due to higher capital liquidity utilisation capabilities.

The new network-based business model has a much stronger focus on the basics of how innovation ultimately creates value and reduces cost both to customers, suppliers and others participants in the innovation process.

A Framework for Analysing a New Global Business Model

Although many researchers and companies today focus on innovation (product, service, process, etc.), not many focus on innovating business models and business platforms. As discovered through the research, only a small number of academic groups who where studying the business model framework[3] were identified. Most of these research groups, however, were merely dealing with clarifying the BM concept from a focal company viewpoint. As a result, it was necessary to define a completely new analytical global and network-based BM framework that could cover the whole chain of customers, suppliers and other network partners.

When developing the framework, previously developed models (Abell 1980, Morris et al. 2003, Osterwalder at al. 2004, Chesbrough 2007) were considered. The object was to find a model that could cover all the elements previously commented on, but none of the models found sufficed.

---

3   For further details please visit the web site – www.NEWGIBM.dk

The new framework suggested here can be seen, both from a single company view point, as well as from a network-based one. Figure 1 below represents each network company's 'open' BM.

Figure 1: Framework for Analysing a New Global Network-Based Business Model

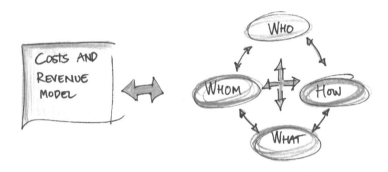

The framework is built on four main research questions:

- Who?
- What?
- How?
- To Whom?

The Four Main Research Questions of the Framework
The first step in using the framework is to describe the exact meaning of the various research questions. The second step is to analyse the relations between the questions.

Who
The answer to the question 'who' should include the different participants in the venture and their affiliation to it. One way of describing them could be to use a stakeholder model. In order to limit the number of interactions in a broad stakeholder model, one could distinguish between the stakeholders who have a close relationship to and participate in the project (like the other partner companies) and those who play a more remote role (like governmental agencies in a regulatory role and other contributors etc).

The purpose of describing the 'who' is, firstly to give the user an overview of the project participants and, secondly, to clarify for later analysis what kind of resources, competencies, capabilities etc. that the participant in question brings to the business venture/network.

What
The question 'what' concerns the product, service or process that the business venture revolves around. In this section, a thorough description of the product, service, process, etc. should be included.

How
The question 'how' can be perceived in two ways. On the one hand, it is concerned with the way the product, service or process will be distributed to the customer, while on the other hand, it also challenges the user to pursue not only incremental innovation improvements, but also radical ones (An appropriate question would be, 'How do you want to innovate the product/process/service etc.'?).

To whom
The question 'whom' concerns the target customer of the product, service or process.

Furthermore, the framework is linked to a cost and revenue model that shows how the new global (and network based) BM is supposed to generate a financial or non financial output, both in the near future and in the long run.

Cost and Revenue Model
The cost structure and revenue model describes how the business venture is supposed to generate financial or non financial revenues in near future and in the long run, and thereby, what the various participants expect to receive in return for their contribution to the business venture.

Interactions Between the Questions
The interaction among the four main questions results in six important analytical areas. The following interactions are not listed after importance. That will entirely depend on the specific case.

Table 3: Interactions between the Questions

	On an Individual Company Level:	On a Network Level:
Who -> What	• Who are the stakeholders that I am planning to work with in order to develop the new product, process or service chosen? (Here, I have an idea (what), but I look for somebody (who) to work with in order to implement it). • What is it that I want to innovate with my chosen stakeholders group? (Here I use my stakeholder groups to generate an idea in the first place) • What is it that I want to innovate? Who do I want to innovate it with? (Here I am very much in the 'dark' – and try to outsource both the idea as well as my network partners).	• Does the network of participants have all the necessary resources (skills, competencies, capabilities etc.) that I need to develop the product, process or service? If not, who should we include in the network? • Does the network have a clear understanding of the value-proposition that we intend to present to the customer?
Who -> To Whom	• Assuming that I have already chosen my customer (the 'whom'), who are the stakeholders that I need to work with in order to best provide for my chosen customers? • Assuming that I do not have yet a target group, who are the stakeholders that I need to work with in order to define my new customer group?	• Does the network of participants have the necessary knowledge about the customer/market that we need to make a successful entry of the product, process or service? If not, how can we get hold of the knowledge? • What are the competitive products? • What is the market size?

	On an Individual Company Level:	On a Network Level:
Who -> How	• Assuming that I have already chosen the 'how', who are the stakeholders needed to assist me with launching the distribution/ innovation of the new product, process or service? • Assuming that I do not know how to distribute or innovate, who are the stakeholders needed to advise me on how to distribute/innovate? • Unlike the conjunction between the 'what' and the 'who' where we concentrated more on what we want to innovate; here we are more concentrated on the level of innovation in terms of radical vs. incremental.	• Does the network of participants have all the necessary resources (skills, competencies, capabilities etc.) that we need to distribute the product, process or service in the manor that we want to?
What -> To Whom	• Assuming that I have already chosen my customer (the 'whom'), what new product, process or service is needed for my chosen customer target group(s)? • Assuming that I do not have a target group yet, for whom do I want to innovate, and what?	• Does the product (distributed by the network-partners) live up to the needs and wants of the customer?
What -> How	• Is the chosen method of distribution/ innovation the appropriate choice for this particular product, process or service?	• Is the chosen network of distribution/ innovation the appropriate choice for this particular product, process or service?

	On an Individual Company Level:	On a Network Level:
To Whom -> How	• Is the chosen method of distribution/ innovation the appropriate choice for this particular customer/market?	• Is the chosen network of distribution/ innovation the appropriate choice for this particular customer/market? • Product development – did it change and how did it change through the process of networking?
Costs and Revenue - model	• Which type of revenue model do I have? • Is the purpose of my revenue model cost efficiency, knowledge gaining or revenues increasing? • Is the purpose of the revenue model long term or short term oriented?	• Which types of revenue do the network partners possess before entering the network formation? • Did the network partners change their revenue perceptions after entering the network? (Both financial and non-financial) • Is the purpose of the network formation cost efficiency, knowledge gaining or revenue increasing? • Is the purpose of the network formation long term or short term revenue model oriented?

Clearly, the transformation from a single BM unit to a network-based BM is transforming individual enterprises' prioritisation. Both the questions asked and the implementation processes are aimed towards ensuring the success of the overall network BM. This inevitably leads to a reciprocal network partners' relation with the belief that 'Their success is also mine, and mine is theirs'.

Conclusion

Most of the innovation literature is concerned with incremental and, especially, radical product innovation taking place within firms and/or in dyadic collaborations. In practice, however, innovation is moving to the level of networks, and consequently the need emerges to tie several firm-specific silos of business development expertise together, create synergies and 'design' a platform to compensate for the deficiencies of individual firms' business models.

The aim of this chapter was to introduce and illustrate a framework of local/global network-based business models. The point of departure in the core elements of firm-level models (Osterwalder et al. 2004).

Later in the book, the authors will show how the framework for analysing new global business models can be put into practice as part of the overall business model innovation process.

Further Research

The new global (and network-based) business model proposed here can, and should be, tested on a larger scale survey possibly supported by more in-depth case studies. For that reason, the authors plan to carry out similar research on an international basis (Europe and USA). The aim is to be able to draw more concrete conclusions regarding the global part of new business models.

References

- Abell, D.F. (1980), Defining the Business: The Starting Point of Strategic Planning, Prentice-Hall, Inc. N.J.

- Barclays/ National Federation of Enterprise Agencies (2002), Profiting from Support, London: Barclays Bank plc.

- Bohn, K & Lindgren, P. (2002), Right Speed in Network Based Product Development and the Relationship to Learning, CIM and CI', CINet, Helsinki.

- Chan, K.W. and R. Mauborgne (2005), Blue Ocean Strategy. How to Create Uncontested Market Space and Make Competition Irrelevant, Boston: Harvard Business School Press.

- Chesborough, H (2007) Open Business Models How to Thrive in the New Innovation Landscape, Harvard Business School.

- Corporate Leadership Council (1995) Vision of the Future: Role of Human Resources in the New Corporate Headquarters, Washington, D.C.

- Kotelnikov, Vadim (2007) New Economy Key Features of the New Rapidly Globalizing and Changing Knowledge Economy- 1000ventures.com.

- Lindgren, P. (2003), Network Based High Speed Product Innovation, Centre of Industrial production, Denmark.

- Lundvall, B.A. (2002), Innovation, Growth and Social Cohesion, Cheltenham, UK.

- Morris, M, M. Schmindehutte and J. Allen (2003), The Entrepreneur's Business Model: Toward a Unified Perspective, Journal of Business Research, 58, 6, pp. 726–735.

- Murphy C. Emmett and Mark. A. Murphy, (2003), Leading at the Edge of Chaos: The 10 Critical Elements for Success in Volatile Times, Prentice Hal.

- Osterwalder, A., Y. Pigneur and L.C. Tucci (2004), Clarifying Business Models: Origins, Present, and Future of the Concept, Communications of AIS, No. 16, pp. 1–25.

- Osterwalder, A. and Y. Pigneur (2004), An Ontology for E-Business Models, University of Lausanne, Switzerland.

- Pasternack B. A. & A. J. Viscio, (1998), The Centerless Corporation: Transforming Your Organization for Growth and Prosperity, Simon and Schuster, New York.

Chapter II.1

Industry Service – Technology Centre

Svend Hollensen

&

Niels N. Grünbaum

Introduction

The Danfoss Group has about 22,000 employees worldwide, factories on four continents, and sales companies and representatives in more than 150 countries. Danfoss has three major divisions: Refrigeration & Air Conditioning, Danfoss Heating and Danfoss Motion Controls. Finally, Danfoss has a large ownership share in one of the world's leading manufacturers and suppliers of mobile hydraulics, Sauer-Danfoss.

Danfoss Industry Service provides internal services for Danfoss' three major divisions. They also provide external services for other non-Danfoss customers (companies), mainly in the local area. The Industry Service - Technology Centre is able to adapt a wide range of services to suit a company's service requirements.

Figure 1: *The Danfoss Industry Service Centre*
Source: IS-T Material

The Industry Service - Technology Centre (see Figure 1) has close relations to all other shared services in Danfoss.

The Technology Centre operates in a locally oriented market where they have a strong position in the southern part of Denmark. They perform numerous services (primarily testing services) in the business to business market for their customers. (For a more thorough elaboration, see below). The Industry Service - Technology Centre is ISO 9001 certified but is not accredited. This, however, has not been a problem, primarily because the centre is a sub-unit of the well known and worldwide acknowledged Danfoss group. The reputation of Danfoss has a rubbing off effect on the Industry Service - Technology Centre, thereby creating an unofficial and unique accreditation that is, indeed, recognised in the market and, perhaps, even more beneficial than the official accreditation. Moreover, this distinctive image is difficult, if not impossible, for competitors to copy and it is inexpensive as well. In fact, it is a differentiation factor that could be one of the most important aspects of the existing business model. That is, it allows the Technology Centre to be able to deliver a high quality product at a very competitive price, thus creating satisfied and loyal customers.

Business Areas

The Industry Service - Technology Centre offers various focused services within the fields of technology, innovation and working environment. Among services performed by the Technology Centre are: technical testing, various analyses (see below), consulting services, calibration, failure analysis, small series and prototype production, and information services. Concerning information services, they monitor and assess competitors' performance. More specifically, testing, analysis and consulting services in Industry Service - Technology constitute:

Testing

- Shock/ bump testing
- Climatia testing
- Mechanical testing
- Vibration testing
- Specialised testing

Analysis

- Chemical analysis
- Metallurgical examination
- Environmental analysis
- Oil analysis
- Surface characterisation
- Wear/ tribology

Consulting Services

- Corrosion
- Cutting fluids
- Supplier assessment
- Materials selection
- Surface treatment
- Product development
- Project management
- Joining processes

Customers

The Technology Centre generated approximately 4 mio. DKK in profit per year in 2005 and 2006. The revenues stem from approximately 70 customers, out of whom, approximately 1/3 are internal customers from within the Danfoss Group. Two characteristics are especially notable in the customer portfolio. All of the customers are located in Denmark and an overweight of customers are located in southern Denmark where the Industry Service - Technology Centre itself is based. This gives a situation with zero internationalisation. This is remarkable, in a period where globalisation is believed to be an excellent opportunity for expanding a given business. Furthermore, the organisation is dominated by a classic engineering culture i.e. a culture that focuses on a mechanical or on very tangible aspects of doing business. This is, for example, reflected in the way employees explain specific working areas and processes. An unfortunate consequence of this organisational cultural trait is a lack of virtual aspects in the business units' services offered to the market. This applies for both the services they offer and for the facilitation possibilities that virtuality represents. In fact, it is fair to say, that virtuality (i.e. information technology, databases, accelerated communication, customer cooperation etc.) seems to be a very promising area when trying to expand the activity level and develop new services for the market, since it is, essentially, not applied in the present business processes.

The empirical data indicated that customers mainly perceived four aspects as important when they do business with the Industry Service - Technology Centre. These aspects are:

1. How quickly the unit can process the requested services (for example failure tests)

2. The employees' ability, time and willingness to listen to the problems that customers present to them
3. The ability to find an efficient and speedy solution to presented customer problems
4. Limited time span between time of payment and documentation of testing results (or whatever kind of service is purchased)

Industry Service – Technology Centre Competitors

When placed on a competitive positioning chart (see Figure 2) the Technology Centre appears in an area where the main focus is on operations support in the value chain (product testing etc.). At the same time, the Industry Service - Technology Centre has oriented itself towards being a total supplier for mechatronic customers. Other companies are more oriented towards taking care of more general business development for the customers.

Figure 2: Industry Service -Technology Centre's Market Profile
Source: IS-T material

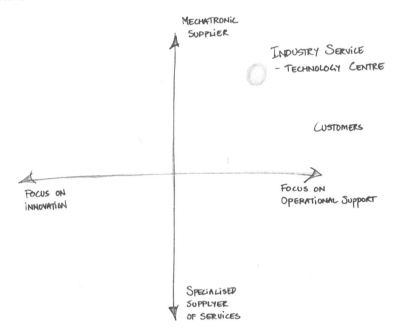

Growth Strategy

The management of the Industry Service - Technology Centre works systematically with strategic scenarios and believes that there is a strong growth potential for the services they provide. They are puzzled about the absence of internationalisation in their activity, i.e. that they only have customers in Denmark and, moreover, that most of the customers are located close to the testing facilities. They believe that different types of barriers (i.e. language, cultural distance, etc.) are important contributing factors. Furthermore, the NEWGIBM researchers have noted that the low degree of market dynamism and the 'mechanical/concrete' mindset of the workers in the Industry Service - Technology organisation, and in the Danfoss group in general, may function as information technology adaptation barriers. The above mentioned elaboration and, more specifically, the absence of internationalisation and of virtual mental constructions have spawned ideas for a new business model. The management believes that this could fuel a development that would allow the Industry Service - Technology Centre to emerge as one of the dominant business actors in the global market. The new business model is described in the next section.

The New Business Model

One of the problems in Industry Service - Technology has been the relatively long waiting time for the testing of a customer product. Reducing this customer waiting time is a key success factor because reducing a product's time-to-market is becoming more important as product life cycles are getting shorter. This was one of the key factors that caused the unit to initiate the innovation project within the NEWGIBM network. Furthermore, the Industry Service - Technology Centre's management considered the possibilities of standardising the services offered to the customers in the form of 'service packages'. This could be achieved by establishing an Industry Service - Technology portal solution, where both buyers and sellers of services have the opportunity to get in contact with each other.

The NEWGIBM project was proposed as follows: The development of an Industry Service -Technology portal solution for the testing of products, including a booking system for standardised 'service packages' in order to reduce waiting time. There are other important elements in the Industry Service - Technology portal as illustrated in Figure 3.

Figure 3: The NEWGIBM Project
– Establishment of a Portal Solution 2007-2008

The important IS-T requirements for the 'portal' have been:

- The portal should provide better Industry Service Technology relationships to both customers and knowledge suppliers and, thereby, make the total value chain of 'product testing' more efficient.
- Customers should be offered standardised 'service packages', which are modularised. This means that the service packages can be adapted to each customer's needs and requirements, based on different standardised service modules.
- Waiting times should be reduced as the built-in booking system automatically searches for 'product test capacity' at the outsourcing partners, if the capacity can not be found at the Industry Service - Technology Centre.
- The portal should offer the Industry Service - Technology Centre the possibility of setting up an auction for generating more 'product testing' capacity.
- The portal should provide immediate access to special knowledge suppliers at foreign universities and special industry testing institutions.
- The portal should provide the Industry Service - Technology Centre with the opportunity to handle more distant international customers and not only customers from Denmark.

The project is scheduled for 2007-08.

Chapter II.2

The KMD Case

René Chester Goduscheit

Jacob Høj Jørgensen

Carsten Bergenholtz

&

Erik S. Rasmussen

The Focal Organisation: KMD A/S

KMD A/S, with an annual turnover of 2.9 billion DKK and 2,800 employees, has traditionally been a company that provides IT solutions for the public sector – primarily the Danish municipalities. For the last few years, however, part of the strategy of KMD has been to enhance the focus on the B2C market in addition to the strong position it already holds on the B2B market. The KMD Energy Division is strongly influenced by the current liberalisation of the Danish energy sector.

History of the Network

The intelligent utility idea and the first outline of the potential business models within intelligent utilities began when the energy area director of KMD visited a seminar for the management group at a business school in Switzerland. An idea generation exercise turned out to be the genesis of the Intelligent Utility Project. The inspiration for the idea was automatic metering of private households and industry, which is currently a growing trend in Denmark and elsewhere in Europe. Automatic metering means that the energy sector can access new and more frequent information about the energy consumption of the energy consumers.

In the period after the manager seminar, the area director did a lot of thinking about the potential of the intelligent utility idea and the potential set-up of an actual project. He shared some thoughts about the concept with some of the customers of KMD. The customers did not immediately buy into the idea of the numerous functionalities of an intelligent utility platform. They merely saw the potential in the automatic metering part as an isolated service. They wanted to make it as easy as possible for the customers to comply with the information needed for the utility providers.

In spite of the lack of immediate interest of the customers, the area director, together with the project manager, was convinced of the substantial potential of the intelligent utility platform. This idea materialised during a 'Blue Ocean' strategy seminar in the summer of 2005, when the intelligent utility concept was analysed as a case. Furthermore, the customers began to express interest in the wider impact of automatic metering and services in this regard.

In the autumn of 2005, KMD decided to let the Intelligent Utility Project become a part of the NEWGIBM Project. By the same token, the area director became aware of the potential of networking with other organisations rather than developing and implementing the intelligent utility network themselves without any external partnerships.

Throughout the autumn and winter 2005, KMD and one of the researchers of NEWGIBM, contacted potential partners in developing and implementing the intelligent utility platform. The project manager of KMD and the researcher arranged meetings with all the organisations in order to present the overall concepts and possible business ideas of intelligent utilities.

An overview of the platform and the data infrastructure outlined before the establishment of the network is presented in Figure 1. The figure was prepared by the second KMD project manager.

Figure 1: Overview of Solution and Data Infrastructure

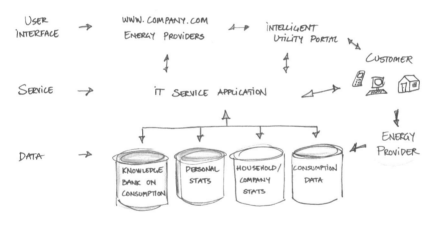

The first network meeting took place in March 2006. This meeting and three of the following network meetings represent crucial occasions in the Intelligent Utility Project and are described below.

Organisations Involved throughout the Period

This section will present the organisations involved in the Intelligent Utility Project at various stages of the network. The presentation provides a short description of the industry in which each organisation is found and gives a profile of the staff that participated at the network meetings.

KMD

The focal organisation KMD A/S brought six employees into the network. The area director was substantially involved throughout the entire project period. As described above, the project manager was deeply involved in the formation stages of the network. However, after getting a new job in the summer 2006, the first project manager, who had a technical background, was replaced by the current project manager, who had a generalist and more business/organisational profile. In addition to the area director and the two project managers, three IT developers participated at various stages of the project.

EnergiMidt

EnergiMidt is a local energy provider in Jutland with 162,000 customers. The company is also engaged in supplying broadband for its customers.

EnergiMidt sent four different employees to the network meetings: a project manager, who was engaged in the implementation of automatic metering (he left the company in the spring 2007) and three other senior managers.

Odense Energi

Odense Energi is a local energy provider on Funen with 72,000 customers, which is the equivalent of 136,000 energy consumers. In addition, the company provides services within electricity construction and other public services.

Odense Energi sent two employees to the network meetings: a senior manager and an employee from their communication department.

NRGI

NRGI provides electricity to about 150,000 customers in the eastern parts of Jutland. The company also provides broadband for its customers.

NRGI sent one employee to the network meetings. This employee was a key account manager with substantial technical insight.

HEF Net

HEF Net A/S provides electricity to about 70,000 customers in the northern parts of Jutland.

HEF Net A/S sent one employee, who was engaged in the implementation of automatic metering.

SEAS NVE

SEAS NVE provides electricity to about 375,000 customers in the western and southern parts of Zealand.

Two SEAS NVE employees from the customer service department participated in a network meeting.

e-Boks

E-Boks A/S provides a one-stop electronic mailbox for both private consumers and for companies. The total number of users has just rounded 1,000,000. E-Boks is partly owned by KMD.

The e-Boks participant at the network meetings was the product manager.

Dong Energy

Dong Energy is a result of a recent merger in the Danish energy sector. The company is engaged in a variety of different services and solutions within energy. The number of customers within sales and distribution of energy and service solutions is 1 million.

Two Dong Energy employees participated in the network meetings: a customer service manager and a technician.

TDC Mobil

TDC provides a long list of services within communication for private households and companies. The total number of customers approximates 11.1 million.

Two TDC Mobil employees participated in the network meetings: A business manager and a representative from the marketing department.

Develco

Develco develops products based on electronics and embedded software for industrial companies. In addition, the company is engaged in automatic metering and management. Two employees participated throughout the network meetings: a sales manager and a sales and marketing coordinator.

Danfoss

Danfoss sells products and services within cooling and heating for industrial customers. The annual turnover in 2006 was 19.4 billion DKK.

One Danfoss employee participated in the network meetings: a director of technical business development.

B&O

B&O offers a wide variety of electronic products. In 2006/2007 B&O had an annual turnover of 4.3 billion DKK.

The B&O representative at the network meetings was a business developer.

Lauritz Knudsen

Lauritz Knudsen produces switches, sockets and plugs. It is part of the global group Schneider Electric SA with an annual turnover of 5.5 billion DKK.

Lauritz Knudsen sent a commercial director and a business developer to the network meeting in which they participated.

Innovation Lab

Innovation Lab is a company employed with trend spotting within technology and pervasive computing. Furthermore, the company arranges seminars with emphasis on idea generation and innovation.

Aalborg University

Aalborg University employs approximately 1,200 faculty and 800 administrative and technical staff. About 14,000 students are enrolled at Aalborg University.

An associate professor (at the start of the project period he was an assistant professor) and a PhD student were involved throughout the project period.

Description of the Network Meetings

Four network meetings have been identified as essential to the Intelligent Utility Project. These meetings are described in this section in terms of participants and content of the meetings. It should be emphasised that these meetings are merely the tip of the iceberg in terms of the network development. A lot of correspondence between the partners of the network has taken place and this has had a significant importance to the network. However, in order to make the description more tangible, the focus of this case description is on the four network meetings that can be seen as illustrative of the remainder of the meetings.

March 2006

This was the first network meeting. Six of the organisations listed above took part in the meeting: KMD (two persons), EnergiMidt (four persons), Develco (two persons), Dong Energy (one person), TDC Mobil (one person) and Danfoss (one person).

The meeting was set up on the basis of the introductory meetings during the winter 2005-2006. The project manager of KMD and a university researcher visited a long list of organisations in order to present the intelligent utilities concepts and to hear whether these organisations were interested in participating in the first network meetings. KMD, together with the university researcher, identified relevant organisations in which to contact. Two companies were contacted but they did not wish to take part in the network meeting.

The researcher made the introduction at the meeting and he lead the discussion together with the project manager of KMD. The discussions were based on unstructured idea generation and brain storming exercises. The outcome of these exercises was a long list of potential products and services that could be provided on an intelligent utility platform. The discussions were also quite focused on the challenges of setting up a project in which all the relevant organisations were involved in combination with the fact that all the participants had a sound business model in this arrangement. The participants committed themselves to meet again on a specific date within a month.

August 2006

Due to different circumstances, (especially the first project manager's resignation from his job) this network meeting was not held within a months' time after the first network meeting but after five months's time. Nine of the organisations mentioned above participated: KMD (two persons), EnergiMidt (four persons), Dong Energy (one person), TDC Mobil (two persons), Develco (two persons),

Danfoss (one person), B&O (one person), Lauritz Knudsen (two persons) and Innovation Lab (three persons).

The area director of KMD decided to use an external event organiser, Innovation Lab, for this network meeting. The intention was to generate new ideas that otherwise would not have been generated. As part of the deal, Innovation Lab contacted a number of the organisations that they determined would be relevant for the network. These organisations were contacted through emails. One of the invited organisations was B&O.

The meeting was organised as an idea generating process with both plenary discussions and group exercises. The first step was to agree on some strategic trends that could serve as a basis for the discussion of specific services. 12 strategic trends (for instance sustainability, customisation, experience economy, etc.) were put forward. Solutions aimed at a wide range of potential services within energy surveillance, consumption management, intelligent metering, smart home, self-sufficiency, etc. were presented throughout the meeting. The meeting was not concluded by an agreement to meet again. Rather, the project manager of KMD stated that KMD planned on continuing the Intelligent Utility Project and that KMD was hoping that some of the participants at the meeting would be interested in taking part in this continued work.

September 2006

At the network meeting in September 2006, three organisations (KMD, EnergiMidt and NRGI) participated. A fourth organisation, Odense Energi was invited but cancelled before the meeting. The meeting took place at EnergiMidt.

The meeting was based on an agreement to move forward with the Intelligent Utility Project. The discussion was aimed at specific problems, such as data security, in the solution provided to the customers. The debate at the meeting was marked by the fact that KMD provided the data handling system, Panda, for EnergiMidt. Thus, the link between the services and products of the Panda system and intelligent utilities was a standing issue at the meeting. In addition, the time pressure of finishing the automatic meter reading solutions for the customers of the two utility providers influenced the discussions to a large extent.

The network meeting concluded in a specific plan for implementing the intelligent utility solutions. It was decided that a project description and a requirement specification should now be prepared for the system developers to follow.

It was agreed upon that the partners should meet again within a short time frame in order to further discuss the way forward.

February 2007

After a meeting in November 2006, which was quite similar to the September 2006 meeting (in terms of participants and issues discussed at the meeting, the next network meeting was set up in February 2007. The number of participants was seven: KMD (two persons), EnergiMidt (one person), Odense Energi (one person), SEAS NVE (two persons), HEF Net (one person) and e-Boks (one person). The participants who were participating for the first time, SEAS NVE and HEF Net, were contacted through the professional networks of the employees of KMD.

The meeting was set up as an idea generation meeting with some creativity exercises. Potential smart home solutions were omitted from the discussions. The KMD project manager lead the discussion and the exercises. It was emphasised that no agreements or contract in relation to the Intelligent Utility Project should be signed as a consequence of the meeting. The project manager described the intention of KMD to apply for external, public funding. Seven partners expressed their interest in working together on this application.

Se *table 1* on the next page for an overview of the four network meetings.

Overall Description of the Outcome of the Network

To sum up the previous section, the network process has generated the vision of a number of functions. These functions can be summed up in the following four points:

1. **Knowledge about energy consumption:** The module would help the customer to get an overview of the current consumption of electricity, water, heat and gas

2. **Energy surveillance and alarming**: This module would provide the costumer with a function that alerts when the consumption of electricity, water, heat or gas is anomalous
3. **Energy guidance:** The guidance function would support the customer in reducing his/her energy consumption through continuous guidance and advice
4. **Smart home services:** This module would be based on integrating all the devices in the private home to one, central control system. The customer could activate and deactivate the different devices from a distance through mobile solutions and/or internet portals. The thought of smart home services is, to a certain extent, based on pervasive computing.

Currently, a consortium consisting of five partners (KMD, e-Boks, EnergiMidt, Odense Energi and a university) is preparing an application for external funding for the Intelligent Utility Platpform. As a point of departure, the platform is based on a subset of the four functions. Thus, only the three first functions will be incorporated into the platform. The latter, smart home services, will not be a part of the platform.

Overview of the Meetings
Tablet 1 outlines the development of the Intelligent Utility Project in terms of participants.

Table 1: The Four Network Meetings and the
Number of Participants at the Meetings

	March 2006	August 2006	September 2006	February 2007
KMD	2	3	2	2
EnergiMidt	4	1	2	1
Odense Energi	-	-	-	1
NRGI	-	-	1	1
HEF Net	-	-	-	1

	March 2006	August 2006	September 2006	February 2007
SEAS NVE	-	-	-	2
e-Boks	-	-	-	1
Dong Energy	1	1	-	-
TDC Mobil	1	2	-	-
Develco	2	2	-	-
Danfoss	1	1	-	-
B&O	-	1	-	-
Lauritz Knudsen	-	2	-	-
Innovation Lab	-	3	-	-

Chapter II.3

Smart House Case

Peter Lindgren

Carsten Bergenholtz

&

Kristin Falck Saghaug

Introduction

Compared to the NEWGIBM setting, the Smart House Network has a different structure. The Smart House Project is not an innovation project that exists in an already given organisation, but a newly started network developed in relation to the NEWGIBM project.

Background of the Smart House Project

Since April 2006, Aalborg University, a number of companies in northern Jutland, business managers and persons from the political environment have cooperated on establishing a business model able to constitute an innovation platform for the building industry in northern Jutland. The network consists of a number of partners, complementary in background and in competences. In total, the network consists of 37 partners, some of whom already knew each other from the building industry and had previously established strong relations, whereas others were not yet acquainted and came from other industries. Some represented the consultancy industry or the scientific world.

Originally, the idea for this network stemmed from an initiative in the Northern Jutland Growth Forum focusing on the creation of a business cluster within the building industry. The origin of the idea was that the building industry seemed to be to be characterised by tough competition in a limited market. The intention of the Growth Forum initiative was to approach this competitive market situation differently, with inspiration from the 'Blue Ocean Strategy' school of thought that focuses on possibilities instead of limitations. Thus, the objective is to create an organisation capable of neutralising the barriers for innovation in the building industry in northern Jutland. It was the objective of the network, that if researchers, companies and users could cooperate on finding the best possible solution - the smartest house possible – a solid foundation for future innovation could be established. The concept was to be exported to a global market, on a long term perspective. The global perspective was to be an integrated part of the concept from the beginning. Furthermore, the objective was not only to focus on the building industry, but to include all other partners in the realisation of the Smart House Project.

The general Smart House concept was not actually new. It is mentioned in articles from 1987[1] concerning the development of houses, including intelligent systems,

---

1   In Futurist sept/oct. 19 87, vol 21, nr. 5 s.52-53

integrating an amount of needs regarding safety, security, comfort, wellness, health and interaction. (A Google search offers 320,000 hits as of June 4th 2007. The concept, however, has never won wide recognition or market-based income. The basic principle is that the available building-oriented technologies provide the opportunity of offering smart and intelligent solutions, combining new or established building projects with the implementation of new technology.

Customer Value and User Innovation

To distinguish itself from other Smart House initiatives, the Smart House Project in northern Jutland, aims to focus on the end user. The project places the customer in the center of the innovation process, and moves towards genuine user-driven innovation. This will position the Smart House Project on an innovative level that bears comparison to huge companies like Apple, Google, Lego etc. This also means a high degree of commitment from users in order to make the Smart House Project capable of keeping development costs low, since the users respond to the products in a way that may give access to new products/innovations.

One of the researchers involved in the project has studied the customer segment in the chain of users that includes other companies, researchers, craftsmen, architects, public institutions, financial investors etc. The value for the different users may be defined as follows:

- End user (residents) - increased satisfaction in daily experiences because of flexibility and the possibility of individual adjustments
- Building industry (producer) - increased business potential because of the sale of unique products
- Public supply undertaking - increased possibilities in remote meter readings of energy consumption and advising on energy consumption
- Researchers – increased possibilities of following the use of new building materials, services etc. in order to document and further develop products and business platforms for this industry

Smart House – Concrete Concepts Depicting the Future Project

Smart House is intended to be a window for exhibiting new ideas and concepts in future innovations within the building industry of the region. The interdisciplinary network of researchers is a vital part of this. The building industry has been eager to develop concrete ideas. After the initial phases, the development of the concepts, during two different workshops carried out in close collaboration between companies, region and researchers gained form. The following five suggestions were developed during these workshops:

1. A physical Smart House - a concrete experimental space
2. A virtual Smart House - to cross the time/space border
3. One or several Smart Houses established in a real user environment
4. A Smart City - to test the ideas on a larger scale, establishing a whole society
5. Smart House within the existing housing stock

On the basis of these suggestions, further innovative solutions were recommended depicting different Smart Houses such as:

- The Pregnant House - the house grows according to the inhabitants' life cycle
- The Earning House - export of surplus energy to neighbours and energy companies (rainwater, solar cell energy etc.)
- Lego - The Flexible House - elements can be exchanged like building blocks
- The Flexible House - colours, structure and lights can be changed

The following list presents a number of components that have been discussed in the Smart House Network:

- Already existing solutions (based on current technologies)
- Supervision of drives
- Movement censoring
- Entrance security
- IT-based control of pools/spas (and of water usage, in general - indoors/outdoors)
- IT-based control of indoor/outdoors lights
- IT-based control of video and sound
- IT-based control of heat
- Ventilation, air-condition etc. (environment friendly)
- IP-telephones
- Wireless services

- Sun-cells in roofing material
- Nanotechnology in windows et cetera

Competitors

The development of Smart Houses is going on world-wide involving huge resources. By building an attractive and unique innovation platform for knowledge and innovation in the region of northern Jutland, the Smart House Project will be of interest to international partners and researchers. By making it very different from other Smart House projects that are carried out worldwide and by avoiding entering the 'Red Ocean', the Smart House Project will have its very own place in the market and, thereby, avoid competition.

Development Process – Partners and Resources

The Smart House Network partially originates from the political environment, but business managers were also involved from the beginning. Structurally, it thus seems that the relevant partners have been involved. Whether the specific people involved have been capable of filling out their respective roles is still to be decided, depending on how the launching of the project will turn out. One of the main problems for the network is the BM issue. Two important questions have to be met with satisfactory answers. First, who is in line to make money and second, when do they want to make it? The first full scale Smart House in northern Jutland will probably not provide any of the partners with a profit, but it has to be seen as a long term project and investment.

The primary force of the network is, probably, that researchers, respected business managers, and top members of the political environment have been included from the beginning. Furthermore, there has been an explicit focus on including users. A problem to overcome here might be a general scepticism in the building industry when it comes to the use of new products. This conservatism has been mentioned from some partners from the building industry.

The Smart House Project has, from the beginning, been an innovation project. The intention of the partners has not been merely to implement known technology. The researchers have also tried to focus on bigger and previously unknown technologies, and new ways to involve different types of users. Since the project has not yet been launched, it is still unclear as to how innovative the project will end up being. This will likely also depend on what kind of funding the future Smart House organisation can procure.

The researchers found that the organisation of the Smart House Network demanded that the future leader and management aimed towards:

- Ensuring innovation in the network across competences.
- Ensuring innovation as 'open innovation' – or focusing on companies and research and development units in the region.
- Ensuring innovation towards commercialisation esp. on the global market.

The Progress and the Future Development of the Smart House Project

Phase 1 activities consist of the development of a BM and the establishing of an organisation for the Smart House Network[2]. These activities (until mid-2007) include:

- Establishing the network
- Establishing the administration
- Publishing/disseminating knowledge

Phase 2 activities are long-term activities. It aims at establishing an entire Smart City with businesses, inhabitants, education, institutions etc. Activities (from mid-late 2007) include:

- Launching projects
- Identifying partners
- Estimating concept potential

In 2007 a manager for The Smart House Project was appointed and the initial steps concerning the development of the Virtual Smart House could begin.

---

2   (Source: "Diskussionsoplæg til møde om realisering af Smart City / Smart House – Nordjylland, 10. april 2007")

For further development and information please see
http://www.smartcitydk.dk and
http://vbn.aau.dk/da/publications/realisering-af-smart-citysmart-house-i-
nordjylland_f5ddbf60-3c55-11dd-a438-000ea68e967b.html

Table 1: Building Blocks for a New Global
Business Model - the Smart House Case

New Global Business Model Building Block	Smart House Case
Value and Perceived Value Proposition	The Smart House concept aims at developing values and perceived values that are both existing and not existing in the industry. Above in the case description some of the new values and perceived values
Process Value and Process Value Proposition	are mentioned. Further the Smart House concept aims at seeing the Smart House in a process perspective which means that values and value proposition will develop along the life cycles of the Smart House.
Target Customer	All network partners target many of the same customers within the same industry. The new Smart House concept will increase their target customer group to customers on the global market and new customer groups outside the building industry. All network partners agree to see this as a possibility to attract other customer groups to the area of interest.
Distribution Channel (physical, digital, virtual)	Both physical, digital and virtual channels.
Relationship	There are both old relations between the partners and some are new and unknown to each other. Relationships to customers, suppliers and other partners within the Smart House concept are developed.
Value and Perceived Value Configuration	All network partners focus on different values and perceived value configuration. These have not yet been classified in details. All network partners agree to see this as a possibility to innovate new values and perceived values to their customers, their product and business model.

New Global Business Model Building Block	Smart House Case
Core Competency	All network partners have different competences. There are still partners from the building industry offering traditional competence to the network.
Partner Network	Physical, digital and virtual partner networks are in the process of being set up
Cost Structure	All network partners have different cost structure and revenue models. The cost structure and revenue models can not be clearly
Revenue Model	defined at the time being. They will be defined spring 2008.

Chapter II.4

The Nano Solar Case

Peter Lindgren

&

Svend Hollensen

The ISO PAINT Nordic A/S Company

Iso Paint Nordic presently employs approximately 30 persons but, due to the acquisition of new production facilities, staff is expected to augment about 50 employees.

The company was founded in the beginning of the 1970s by Joen Magnus Reinert. The managerial responsibility has now passed into the hands of Managing Director Nina Reinert and Development Manager Joen Reinert. Gradually, however, the entire managerial responsibility will be passed over to Nina Reinert, thus ensuring an alternation of generations.

Iso Paint produces roof coatings, impregnations, facade painting as well as specialised products for surface treatments.

Iso Paint has specialised in the production of paint for special purposes in modern production facilities. The company owns its own research laboratory for the purpose of developing new and environment friendly products. In 2003, Iso Paint was awarded the EU Eco-label 'The Flower' for one of its products.

The Market

Technologically speaking, Iso Paint works within the field of well known and well established technologies but with a constant focus on development of production techniques and of the components added to the paint.

Iso Paint is certified according to the ISO 9001:2000 standards, which provide assurance about the ability to satisfy quality requirements and to enhance customer satisfaction in supplier-customer relationships. Iso paint is also certified according to the ISO 14001:2004 standards defining the requirements for global environmental management systems and for organisations wishing to operate in an environmentally sustainable manner.

The innovativeness of the industry is high on all major markets. The markets are characterised by fierce competition on product, marketing, price and, in particular, on distribution. Many innovation initiatives are launched in relation to distribution channels and management of distribution channels – for example, third party logistics providers. There is an increasing tendency to skip elements in the supply chain, when possible, and to take over the stocks of the retailers. All the important retailers, e.g. the DIY centres establish chains of stores that cooperate with each other, thus growing large enough to allow them to take over the role as wholesale distributor. Likewise, an increasing number of wholesalers establish their own production so they can offer their customers a total solution and a broader product assortment.

Several competitors in this field have launched a 'direct' concept that allows them to reach the customers directly by means of their own stores or by taking over vending space for the customer. This is the case for companies like Bilka and Silvan.

Iso Paint does not, however, concentrate their efforts on the traditional retailing sector but, over the previous years, they have established their own network of roof painters, floor grinders and other workmen using Iso paint products.

The network has been built up by the establishment of personal contacts and yearly agreements with each customer.

Presently, Iso Paint does not have a large share of the market for paint in Europe. However, the company has a comfortable share of the market for special paints, especially roof paints and surface products. In Denmark Iso Paint is the market leader. Iso Paint markets itself on a carefully chosen number of European markets. Germany is by far the largest export market.

Figure 1 illustrates Iso paint's business model in relation to three factors:

- Customer groups serviced
- Customer needs satisfied
- Technologies used to satisfy customer needs

The complexity of customer groups, customer needs and technology will increase as the box moves out / up on the axes.

The analysis shows that Iso Paint presently satisfies customer needs for surface treatments of most types of material. The customers are constituted by a large spectrum of B2C and B2B customers. Customer needs are found on the y-axis from the inner level in the less complex zone to the medium level of the axis where the more complex products are to be found. Paints and surface treatment products are primarily sold to business clients and, to a smaller extent, to private consumers via the Internet or in a circuit of distributors and own shops.

Figure 1: Abells Model

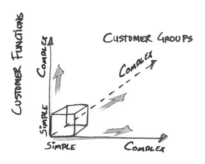

The buying processes of the business to business clients are relatively simple because the number of participants in the buying process is limited, the buying process is known and the after sales process is predictable. This means that most customer groups have a very low degree of complexity and they will be found at the beginning of the z-axis.

Some of the customer groups, however, are more complex and show a complex buying behaviour. These groups comprise e.g. municipalities, regions and the state. These groups are very interesting in Iso Paint's perspective.

Iso Paint relies on standard knowledge and technology on how to mix and produce paint to satisfy the customer needs. In the future, however, the basis of the BM will move further out on the x-axis as more complicated and less known product and production technology will be integrated in the development of new paints. This could include the use of new components or a combination of components with new techniques such as nanotechnology.

Technology, Market and Innovation in the Future

Iso Paint will adapt new technologies (new materials, new production technologies and an extended product integrating more immaterial features in it). E-Business will become a main focus issue. The markets will be internationalised and relations to new customer groups will be established. The innovation degree will be located between incremental and radical innovation.

Figure 2: Balachandran's Contextual Cube Model

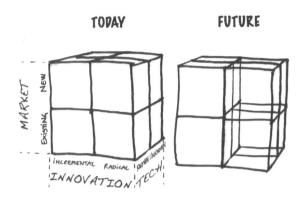

Figure 3: Abell's model illustrating Iso Paint's situation

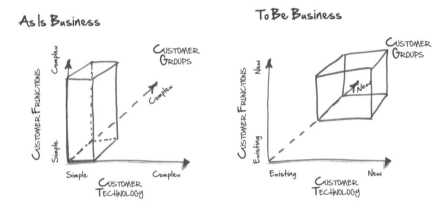

Iso Paint's situation in relation to customer needs, customer groups and customer technologies is illustrated in Abell's model below:

Customer Needs, Customer Groups and Customer Technology Today

The analysis has demonstrated that Iso Paint presently satisfies the relatively simple customer need for paint for buildings, traffic separators and roofs. It is, therefore, easy for the customers to define their needs and the customer needs are, therefore, found at the beginning of the y-axis in the less complex zone.

The paint is primarily sold to B2B customers and a circuit of retailers. The buying processes of the private end-users and of the B2B clients are relatively simple because the number of participants in the buying process is limited, the buying process is known and the after sales process is predictable. This means that most customer groups have a very low degree of complexity and they will be found at the beginning of the z-axis.

The technology used to satisfy the customer needs is also relatively simple because the technology is known and involves the use of standard components – paint components that are mixed to known standard and special products according to different recipes that have been developed internally in the company.

In the future, however, the basis of the BM will move further out on the x-axis as more complicated and less known product and production technology will be integrated in the development of new paints. This could be achieved through the use of new components or a combination of new components and new techniques such as nanotechnology.

Customer Needs, Customer Groups and Customer Technology in the Future

Iso Paint is presently aiming at a development that will ensure the capacity of the company to handle an increased number of markets with different customer needs. Iso Paint must be able to handle a higher degree of adaptation to customers' needs within the field of paint, distribution etc. In the future they must also be able to handle features like delivery, service and documentation on the web. This means that Iso Paint,, has to develop competencies allowing them to produce web based services and products and the company will depend more on its ability to do so.

Porters 'Five Forces' model is used to analyse the competition on the market. 'Rivalry between Competitors in the Industry' is found in the centre of the model and the four forces that influence this rivalisation surround it. This shows how the rivalisation within the paint industry is today and how it will be in the future. Iso Paint is presently in a competitive situation characterised by a strong buyer position and where threats of entry from new competitors on the market are impending. The existing competition on the market is tense with a strong focus on price and quality. The suppliers also hold a strong position and the threat

Figure 4: Porters 'Five Forces'

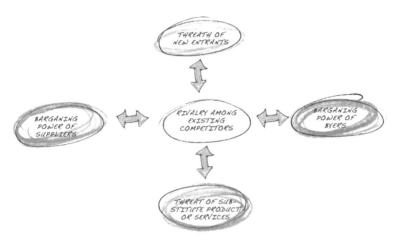

of substitute products is impending. Entry barriers are relatively insignificant, whereas exit barriers are high because heavy investments in production facilities are required. Therefore, the industry is characterised by a strong rivalry. In the future, the power of the customers will increase. The threat constituted by the possible entry of new competitors is expected to increase considerably, thus enhancing the competition on the market. The suppliers are also expected to strengthen their position and are expected to send new product categories on the market that are likely to substitute some of the existing product in Iso Paint's present assortment.

Finally, an extra element has been added to evaluate the market situation in order to show the dynamics of the market. It is the three alternative definitions listed by Ron Sanchez.

- The first market situation is characterised by 'a market with stable technologies and stable market preferences - the result being a focus on the costs associated with the production of standardised products. Product development is periodic and rare.'
- he second market type is characterised by 'changing or developing market technologies and preferences. It becomes a question of how to exploit the technologies to diversify products on the market from those of other companies. Focus on cost is still great but not crucial.'
- The third market type is characterised by 'a market where the technologies change rapidly and frequently. The company focuses on rapid integration of

Figure 5: Ron Sanchez

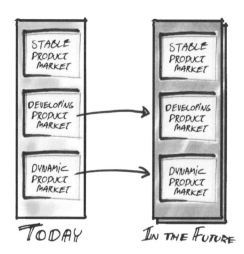

new technologies so the company will be able to face much more sophisticated customer needs and wishes. Product replacement on the market is high as it is vital to continually introduce still better products.'

Iso Paint's product and market situation is undergoing a strong development. To a certain extent it can be characterised as a dynamic development. This is also expected on a long-term perspective. The price parameter is expected to become even more significant on the market. The same will be case for the technological and performance aspects of the products. It is also expected that the company, thanks to the technological development, will be able to target new customer segments.

Product Development

The company has constantly had its focus on production as well as on product development. The latter has always played a vital role in the company. To keep pace with the growing demand from customers of better, cheaper and different types of products, the pressure on product development has increased too. Due to a very committed product developer, Iso Paint has, however, been able to occupy a leading position on the market in relation to competitors. The development activities are carried out and managed by Joen Magnus Reinert in cooperation with a new manager of the lab facilities. At present, Iso Paint focuses primarily on an ongoing improvement of products and production, whereas the radical product development is carried out in shorter, well-defined projects and within the field of products that already figure into Iso Paint's product portfolio but are based on new technologies and materials. The sales and marketing teams have started development projects on the specific markets. Generally, Iso Paint has chosen to carry out its different development projects as 100 percent in house activities although Iso Paint, in certain situations, has relied on cooperation with subcontractors.

Besides Joen Magnus Reinert, the development team consists of 2–3 employees from laboratory. The development activities can not be described by means of a Cooper Model as they are primarily carried out as small inventions made by Joen Magnus Reinert. It must be stressed that a consequence of the new ISO9001 certification is that the company has been faced with the necessity to formalise its product development procedures more.

Ideas for new products are generated from the market but also from internal sources, e.g. Joen Magnus Reinert.

Figure 6: Coopers Stage Gate Model

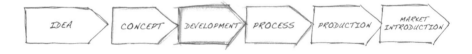

Competences and Core Competences

Which technology, innovation and market competencies does the company possess? Are they company or person specific?

Technology

The primary competence of the company in this field is to mix chemical liquids according to prescriptive formulas. Today the production is organised as a combination of automated paint production lines and the manual mixing of smaller quantities of paint and varnish for special purposes.

Network

Iso Paint presently participates in a number of informal networks with developers from other companies within the paint industry. It is Iso Paint's hope that this group will be able to detect budding possibilities and challenges for the company.

The Innovation Project – the New Global Business Model

ISO PAINT NORDIC A/S is considering taking up new business activities, thus renewing the old business models. Particularly, nano solar painting was in focus. The research and development department came up with the following ideas and new business possibilities:

1. Nanopaint for protection of all outdoor surfaces against:
   - General degradation – less sun, 'the light turns surfaces green'
   - Algae, mosses and green growth on all external surfaces
2. Nanopaint for protection of all indoor surfaces against:
   - Mildew inside buildings, e.g. schools – this can cause disease
3. Against UV light from the sun – degrades everything
4. Nanopaint for the production of friction free surfaces (Cars, windmill blades, underwater paint for ships and pools)
5. Nanopaint to augment the resistance and reduce the propensity for dirt
6. Nanotechnology in paint to absorb the sun and the daylight in order to liberate it in the darkness (Traffic separators, crash barriers, escape route indicators)
7. Roof coatings and façade paints with built-in solar cells based on special nanotechnology which is able to absorb and transmit energy in the form of heat or electricity

All these new possibilities would demand enormous resources of Iso Paint Nordic, both in terms of manpower, lab equipment (electron microscopes) and special production facilities. The total project production would be extremely cost intensive for the company.

Iso Paint decided that the final NEWGIBM Project should be: The development of a nano solar ICT based project having the capacity to change the whole cost structure of a customised project in 3 areas:

1. Building area – façade renovation in Eastern Europe
2. Roof coating – the global market wherever the sun is a possible source of energy
3. Production facilities

Firstly, when implemented, the benefits will be as described above. Secondly, the customers will benefit from huge energy savings, energy stability and security. Thirdly, the payback time on the investment could be diminished tremendously for the customer. Fourthly, the incentive to make 'façade renovation' will increase as it will also be an investment in a future energy supply.

The innovation project and the new BM are focused on the global market and intended to be developed together with Danish researchers and companies (Aalborg University, Iso Paint, Power Lynch, Danfoss), German researchers (University in Magdeburg – Farbe und Lak Institute and Farbindustrie, Dresden) and 3 customers/partners in Serbia, Asia and Australia.

The project is scheduled for 2007-08.

Chapter II.5

The Master Cat Case

Peter Lindgren

Yariv Taran

&

Jacob Høj Jørgensen

Introduction

A Norwegian and a Danish consortium of companies and organisations decided to develop a new business platform – a new 'Ryanair on water'. The Master Cat vision was to develop a new high speed ferry (1 hour and 45 min. as opposed to the present 3 hours and 45 min.) from Kristiansand (Norway) to Hanstholm (northwest Denmark) at only 30% of the normal market price (DK 975, - for 1 car and 7 persons compared to a normal price DK 2995,- for 1 car and 5 persons).

The new idea was a consequence of a mature ferry market, characterised by a small number of players and high prices. The market looked basically like the air flight industry before the low price carriers entered the market. This is the reason why the Master Cat project was labelled the new 'Ryanair on water'.

The Danish and the Norwegian Networks

The local network formation ties and partnerships constituting the Master Cat network, however, were very different from one another. On the Danish side, the network formation was dominated by individual network partner relations, who had decided freely and voluntarily to enter the network. Some of these partners had worked with each other in the past, while others were new to the network. On the Norwegian side, the network formation was much more formalised (joint venture). An agreement between the various partners was signed, in order to ensure a mutual accomplishment of the various business objectives.

Due to the diversification within the network, formation ties and partnerships (in each country), individual business models, as well as individual prioritisations, changed respectively too. The primary focus of the Norwegian partners was to establish a profitable business – a new fast and low-priced ferry between Norway and Denmark. On the Danish side, although some partners were also focused on the same goal, others were more interested in exploiting the new creation as a stepping stone to develop the entire region of Hanstholm, particularly with emphasis on a better exploitation of the tourism potential, which is expected to grow significantly in volume.

The Importance of the Nature of the Network Configuration

The Master Cat case illustrates the simple fact that the nature of the network configuration can determine, to a large extent, the individual network partner's business model value equation. In a more 'contract-free' network, the values and BMs set by each participant are expected to be more flexible or diverse.

In the Master Cat case, the network formation ties can be considered local as well as regional (Denmark-Norway), and the new ferry can be categorised as a new product/service to the region. The more the network partners are locked-in (joint venture) into the network-based BM, the more committed they will be to ensure the overall success of the specific task chosen and, therefore, individual partners will adjust their BM formation to better fit in with the overall network-based BM structure.

In the following table the Master Cat case is related to the building blocks of the New Global Business Model:

Table 1: Building Blocks of the New Global Business Model

New Global Business Model Building Block	The Master Cat Case
Value and Perceived Value Proposition	Norwegian Group: Establish a profitable ferry business Danish Group: Establish a profitable ferry and tourism business to develop the entire Hanstholm region
Process Value and Process Value Proposition	
Target Customer	Both groups share the same target customer – ferry passengers and cargo customers The Hanstholm group also see this as a possibility to attract other customer groups to the area – skiing tourists, other cargo companies e.g.
Distribution Channel (Physical, Digital, Virtual)	Physical Channel
Relationship	Norwegian Group: Customers are the passengers on the ferry and cargo customers Danish Group: Customers are both 'on' and 'off of' the ferry
Value and Perceived Value Configuration	Norwegian Group: Share common values Danish Group: Diversification of values and perceptions

New Global Business Model Building Block	The Master Cat Case
Core Competence	Norwegian Group: Strong network ties through contracts (venture capital), harbour, resources Danish Group: Strong network ties through trust, harbour, resources
Partner Network	Local/ regional network
Cost Structure	Short and long-term financial focus
Revenue Model	

Concluding Remarks

Seen in a product innovation perspective, the Master Cat innovation project can be considered incremental, judged on the technical, market and network dimensions whereas the business model can be seen from two angles: In the region of Hanstholm (Denmark) the authors consider the network-based BM to be a 'realisation model', since the change that is required by the participants to form the network-based BM is relatively small. For the Norwegian group in Kristiansand, the change is considered to be a 'renewal model'[1], since all partners signed a contract - venture capital, and, therefore, they were more 'locked-in' to the new network-based BM formation.

The Master Cat case was also the beginning of a much stronger collaboration between the different companies present in the Port of Hanstholm and Aalborg University. This collaboration focused on innovation. Through a number of innovation workshops, the Hanstholm consortium came up with the following 5 business areas to further develop.

1. Fish
2. Cargo
    A. Cargo – Fresh fish
    B. Cargo - Gravel, Raw Materials
    C. Cargo - Containers

---

1   Linder and Cantrell 2000

3. Ferries
4. Tourism
5. Service Industry

The consortium decided that all these 5 areas should be innovated in separate projects but that the individual innovation projects should be developed on a shared innovation platform. The consortium, therefore, decided to develop a new innovation platform called Smart Harbour – The Intelligent Harbour. In 2010 a new business model project regarding wave energy came up.

The companies desired to develop new business models that applied to Hanstholm and their industries because the situation in all 5 existing business areas could be characterised as 'Red Ocean'. The companies behind the platform considered this situation to be even more 'Red Ocean' because of a high concentration of customers, larger customers and more professional global customers. Furthermore, a number of competitors in some of the business areas invested heavily in production capacity, cost efficiency and, thereby, cost leadership.

For more information and latest update please see:

- www.hanstholmhavn.dk/en/port_of_hanstholm/port_of_hanstholm.htm
- www.hanstholmhavn.dk/dk/hanstholm_havneforum/

Chapter III.1

The Pitfalls Of The Blue Ocean Strategy – Implications Of "The Six Paths Framework"

Peter Lindgren

Kristin Falck Saghaug

&

Suberia Clemmensen

Abstract

The Blue Ocean Strategy (Kim & Mauborgne 2005) has been one of the most important analytical techniques related to the area of innovation and new business model (BM) innovation since 2005. Today many consultancies and managers responsible for innovation use the Blue Ocean framework as one of their top 5 innovation tools. Addressing the Blue Ocean framworks's tools, this chapter accentuate the importance of using the Six Paths framework carefully when generating new business models. It is important to understand The Six Paths frame work's foundation – value. This chapter address the questions: What does it mean to innovate values related to the Six Paths framework? and What are the implications, challenges and pitfalls of the Blue Ocean Strategy's Six Paths frame work – related to exploring and developing the values inherent in the company, when pursuing to create a new market?

Keywords: Blue Ocean Six Paths framework program, Value, Creativity, Innovation of Business models, Business Models.

Background

Numerous authors have developed a list of tactics and analytical techniques to discover new business models (Markides 2008, Osterwalder 2009). The Blue Ocean Strategy (Chan Kim & Mauborgne 2005) has been one of these - probably one of the most important analytical innovation techniques to new business model (BM) innovation since 2005. Though have there been voices within academia and consultancy that have been less enthusiastic. Some are claiming that it is just an incomplete strategic consultancy tool (www.trugroup.com)), or a way to create temporary monopolies, (Lele 2005). The consultancy vSente's managing director Mike Smock claims that the creation of a blue ocean is just another way to attract more competition (www.twoscenarios.typepad.com), and another academic article argues that it is risky and problematic for the company to simply leave their investments and value systems (Christensen 2006). Our aim is not in particular to continue this critique here, but to address the several implications we found during our research by testing and using the Six Paths framework related to value innovation.

"Value innovation is a new way of thinking about and executing strategy that results in the creation of a blue ocean and a break from the competition." (p 13, Chan Kim & Mauborgne 2005).Wanting to belittle the importance of value innovation, when generating new BMs - it is important to understand the very foundation - value - in the Blue Ocean framework – both to the Strategy Canvas, the Six Paths framework, the Four Phase Matrix and finally the Blue Ocean Strategy formulation.

The Six Paths framework conceptualization of values and the process of finding and looking for new values are aimed at establishing and eventually create the new value curve (Kim & Mauborgne 2005). New values and variety of values (Alderson 1957; Drucker 1973; Albretcht 1992; Anderson 1982; Woodruff, 1997; Anderson 1999; Doyle 2000; Wouters 2005, Osterwalder 2005) are also related to the understanding of the implications, challenges and pitfalls of the Six Paths framework. The reason for this is that value are strongly inter-related to the relationships between the customers, the market, the industry, the network partners (Blois 2004) and last but not least the company itself.

The Blue Ocean Six Paths framework concept will according to Chan Kim and Mauborgne lead to develop new values and thereby BMs with real market and customer values and opportunities. The purpose of this paper is to examine how the Six Paths frame work concept related to value and perceived values was applied to value innovation of new BMs in companies and where the implications, challenges and pitfalls of the framework were found related to defining value.

Research and Findings

This paper is pursuing some preliminary research into the area of the tools related to the concept of Blue Ocean Strategy; focusing especially in this paper on understanding and analyzing the implications, challenges and the pitfalls of the Six Paths frame work program and the value curves. – to be continued in the following research and paper with the Four Phase Matrix and the strategy concept – focus, differentiation and unique point Previously we have analysed the pitfalls of the strategy canvas addressing the implication, challenges and pitfalls of defining and using value related to Kim and Maubourgnes strategy canvas framework.

Research Question

The main research questions address:

What does it mean to innovate values related to the Six Paths framework program?

What are the implications, challenges and pitfalls of the Blue Ocean Strategy's 6 Path frame work – related to exploring and developing the values inherent in the company when pursuing to create a new market?

Design/Methodology/Approach:

Case studies of small- and medium-sized companies involved in BM innovation provide the empirical basis.

The study was conducted from 2005 to 2008 as a part of the New Gibm project and Blue Ocean project funded by the Danish Ministry of Innovation and Science and Danish Ministry of Economic and The European Commission. This paper is based on case studies and the approach is action based qualitative research (Sharmer 2009, O Brian 2007) and participative research (Wadsworth, Y 1998) inspired by Paulo Freire related to development and learning in Fals-Borda, O., & Rahman, M. A. (Eds.). (1991). In accordance with the founder of action research this approach is not inferior to "pure science" (Lewin 1946) and in (Godushecit 2009).

The approach is cross disciplinary and explorative as very little research in this field was available at the start of the study. Researchers were involved from the first workshop's inception and took active part in the SME's work with the blue ocean framework. Furthermore, the companies participated as group members in the networks of SMEs and took part in the discussions, development and screening of values related to the Six Paths frame work program focused at innovating BMs. This was done both at individual meetings with SMEs, SME partners and at group meetings. Notes have been taken from every interview and meeting or they were taped and transcribed into files and uploaded to the project extranet. We filed the data collected for each of the companies, using the following broad categories:

* Individual partner data
* Information about the individual SME's strategic canvas, 6-paths, matrix and proposal for new business models
* Key aspects of the strategic canvas, and the BO implementation process

The Six Paths Framework

"The Six Paths framework" is introduced as the second tool after "The Strategy Canvas" in the Blue Ocean framework. The Six Paths frame work consists of 6 "paths" to find new tangible and intangible values. The tool is directed towards a redefinition of "the market" to the company and it is meant to achieve a perspective that looks beyond the current limits of the company and its existing markets. These are the 6 perspectives the the Six Paths frameworks adresses:

1. Alternative industries – instead of competing within the industry
2. Strategic groups – instead of established strategic groups
3. The chain of buyers – instead of narrowing focus to the same buyer group
4. The complementary products and services - instead of the industry's scope
5. The functional and emotional orientation of an industry – instead of same orientation as industry
6. The time scope – instead of addressing the same time as industry

In many sense it is a radical tool for innovation and it is a tool for "different thinking" - a concept that may be understood as an ability to balance between one's own individuality and at the same time be participating (Saghaug and Lindgren 2009). Inspired by a more philosophical and theological approach based on thoughts regarding human existence and ability to create originating from Paul Tillich(1951) – one needs to make relations into the unknown – into strangeness – to find potentials for innovation. These relations may be perceived as both strange, in the sense unknown but also as peculiar or odd. As such it is a matter of open innovation where the company in order to address the unknown future is relating to for instance knowledgeresources outside the company in order to develop and innovate its own BM (Chesbrough 2006).

Values are not a neutral concept and the understanding of the metaphor value is multiple related to context and paradigms (McCloskey 2001). In a former paper on elaboration of the Blue Ocean tool the Strategy canvas, a number of approaches were provided (Lindgren & Saghaug).

This paper continues the work in addressing what sort of values the different paths in the Blue Ocean framework may refer too. This is done because the Blue Ocean book, though focused on values, aims through the Six Paths at value innovation, but do not define what sort of value the different number of paths may address. The paper's hypothesis is that The Six Paths operates with values on several levels, but they are not explicitly formulated. The point of entry to this analysis and each of the Six Paths is the mindset of the framework and the value mindset of the Blue Ocean theory – the perception of values that the path originally was meant to deal with. Secondly a description and illustration of how the path is defined, thirdly what sort of value this addresses and then case reflections and findings which enlighten the path and its implications is referred. Finally a preliminary conclusion related to the specific path and its value proposition is elaborated.

The 1st Path - Alternative Industries – Instead of Competing Within the Industry

Mindset: The mindset in 1st path of the Blue Ocean framework is according to the authors not to list rivals and focus on rivals in the same industry as the company currently is placed and to compete with them. Focus is instead on alternative industries, their value offerings and which values that can be transformed to our industry.

Definition: The company in focus are motivated to look into several alternative industries and analyse their products and services which often have different functions and forms. However they often serve the same purpose according to the Blue Ocean theory and therefore the same value fulfillment. One Blue Ocean example is the Cinemas industry and the restaurants industry serving the same purpose and value but with different form and function - enjoying a night out. Further case examples from the Blue Ocean book are PCA (financial software packages) and NetJet (partial aircraft ownership).

The participants dealing with this 1st path are in other words asked to focus on alternative industries' values. This method should open their eyes to value-offerings and alternative value offerings and thereby gain new horizons and find new key values that can either attract new customer groups or improve the value proposition in the existing industry.

The questions that Six Paths framework asks are firstly: What are the alternative industries to your industry offering – and which values exists in alternative industries. Secondly it questions: Why are customers trading across alternative industries? By this the participants are focusing on which values attract customers to trade up- and down in alternative industries to fulfill their needs (Kotler 2009).

Case - reflections: Basically the value systems within two alternative industries are normally different. This research showed that the concrete and tile industry had two very different value propositions. Consequently core values of each alternative industry and the context these values are placed in become very important to understand and know before they are transformed into new ideas in an alternative industry. What values may be added to your existing customer's demand and how is the value in focus related to other values? How will one value be related to existing values in this industry? Tapestry vs. raw concrete wall was another example in this context because values related to tapestry industry was related to design, form, fashion and possibility of changing easy, but concrete walls was related to easy building, cost efficiency of building, sustainability and robustness.

To have both a short and longterm perception of the situation for the customer becomes important, as well as the aim for the innovation. Tapestry has to be change more often than raw concrete walls – opposite tapestry industry will have some difficulties.

As well as the current value system in the concrete case company at hand – e.g. the understanding of leasing tiles – meaning combining a profit formula from another industry – e.g. car leasing - with tiles was extremely difficult to the company to comprehend and find meaningful.

So in order to succeed with this path the researchers was constantly challenged related to how far the company was able and willing to think on the basis of their core values and existing business model. The core of the values becomes necessary to understand, because it consists of many different values that influences each other and influences the way the company perceives values in their industry. The time scope is also not taken into consideration as the case findings showed that values change over time and further the priority of values differs over time.

Another issue is what is an alternative industry really - an extremely relevant question to discuss. Which values is one competing on in an alternative industry (again impossible to generalize but important to address).

The 1th path values deals with what we have illustrated in a horizontal industry value set dimension.

One industry is looking across industry boundaries into value sets in other industries.

Values addressed in alternative industries and with different form and functions addressed as if they were compatible with each other.

Figure 1: The horizontal industry value set dimension

Preliminary conclusion 1: The 1st path requires an extremely high degree of open-mindedness of the company and an ability to "play" with a huge value leap. The cases showed that values are not just values. The content of values as for instance related to an evening "down town" may be based on very different values, value sets and consequently be very difficult to generalize as covering a whole group. One needs to address a diversity of customer groups and their values if this path should be relevant and give fruitful results and new ideas for new markets. Further time also plays an important role to values and which values can be transferred.

The 2nd Path - Strategic Groups – Instead of Established Strategic Groups

Mindset: The 2nd path of the Blue Ocean framework focuses not on competing positions within the same strategic group, but on finding new options addressing different strategic groups within the same industry.

Definition: By addressing these strategic groups within the same industry the idea is to see which of the companies in the industry are possibly following a similar strategy and performance. The focus is to try to make companies break out by understanding which premises and "values" that influence the customers' decisions for trading up or down from different strategic groups in the same industry. Which customers' demands are served and which are not served in each strategic group.

In this context the values are looked into on what we call a vertical strategic group value set dimension.

In this context one needs to understand where new and unknown combination of values from different strategic groups may be made. In the Blue Ocean book the Curves case are mentioned (cheap and easy accessible fitness only for women) combining customer groups found at home with values as the privacy of a training video and a fitness center as a social forum. Another example is Lexus Toyota

Figure 2: The vertical value dimension

that had the luxury looks of a sport car outside but the price and cost of a family car. Thereby these companies provided the possibility for customers normally frequenting another strategic group to trade across strategic groups.

The questions to be asked in this context are where and what are the strategic groups within your industry? Why are customers trading up or down between groups? Which values are they trading on?

Case reflections: The focus in the case research was on strategic groups and customer groups within the same industry and what kind of values they were trading on behalf of. The strategic industry's focus and mindset of values needs in this context to be very observant - which means moving from one value position to another value position but on a vertical dimension. The question is however, does these value and procesual value move mean anything to the company? The experience from research was as in the 1st path that another strategic group's attitudes and convictions – understood here as values – firstly mean something to the company if they are meaningful related the company's own values. Otherwise they cannot or will not adapt these values. They will not act on it.

Within one case - the fishery harbor - one of the attendant claimed, related to his perception of his industry's customer demand, that "the only thing that matters is price, price, price!" Further the research discovered some hidden groups related to the chain of customer and the life cycle of the product and service which influenced the values and the priority of values. In the "Concrete company case" example the customers were often not the decision-maker and further the products put before the customers house had to be recycled and taken care of – which suddenly became a task – a value - to the company to solve and fullfill.

The vertical strategic group value set dimension might also be influenced by horisontal and not to the company visible values. A vertical value dimension – a value proposition from another strategic group – or a horizontal value set dimension – a value proposition from an alternative industry - may push the company to adapt new values or the opposite wait or deny adaptation.

Values addressed by horizontal and vertical positions can therefore be highly attractive to a company but influence the final result of the value output to the customer – the buyer.

Figure 3: Vertical strategic groups value vs. horizontal alternative industries value

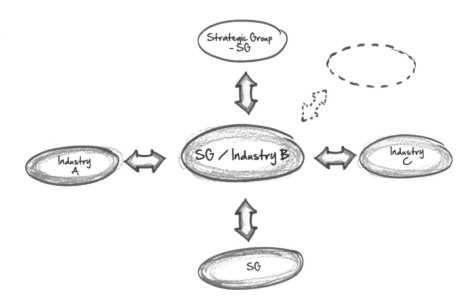

Preliminary conclusion 2: One challenge is the perception of the vertical strategic group value proposition. To address who might the strategic group might be, and what is the ability at the company to interpret these values, address and act on them (Saghaug & Lindgren 2009). If the values are very different form the company the establishing of a relation will prove to be very difficult. Another dimension is the "hidden values" in the chain of customers which is explained beneath but cannot be explained by the horizontal value set and vertical strategic group value dimension.

The 3rd Path - The Chain of Customers – Instead of Narrowing Focus to the Same Customer Group

Mindset: Direct competition is focusing on existing customer but the Blue Ocean framework tries with this 3 path to move the focus to new and non to the company existing customers – in the chain of customers.

Definition: The 3rd path sets out to redefine the industry and to create new customer segments. This is done by focusing on the chain of customers as figure depicts

Figure 4: *The chain of customers*

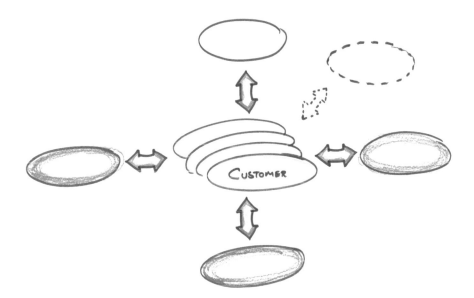

The companies within one industry often address the same customer segments, for instance large customer groups vs. small customer groups. Firstly the path tries to focus on the values along the whole chain of customers in order to find new potential customer groups. The case example from the Blue Ocean book is Novo - who according to the authors invented the insulin pen by looking at the end-user and not the physician. The questions to be asked here are - What is the chain of customers in the industry? Which groups of customers is your industry traditionally focusing on? And If you frequented down and upstream within the customer groups of your industry, how could you then open up to new values (new demands) and new customer groups?

Case reflections: The Handicraft case showed that the values of another customer group were very different to the existing customer groups' values. The values of different customer groups are seldom comparable and further their priority may change. Further some values do not even exist or are prioritized very low in the mindset of the customer group in focus.

Values addressed in this context are unknown, but they are also related to products/services and buying situations in quite a different context.

Preliminary conclusion 3: The 3rd path touches upon some values quite different to the values in the 1st and 2nd path as this framework moves outside the primary customer group to the industry and focus on those customer that are before and after. It follows the product and service through the chain of customers, which is often very difficult - also in relations to values and meaningfulness to the company. Though how are values developed through the chain of customers and how is it influenced by time? The understanding of value for customer in the chain of customer groups will according to our research be very different from the companies' existing customer groups and turned out often to have another meaning or priority. Also when focusing on values in another part of the customer chain, the context in which the value is operating, is important to understand. Therefore it is once again important to address that the interpretation of values for the different customer groups in the chain and the concrete context the value is operating and used in - is very critical to understand. Another important issue to address concerning the innovation process is how this can be related and transformed to the values of customer in the Strategy canvas later on. To be accurate it would need to be much more complex than the Blue Ocean authors show.

The 4th Path - The Complementary Products and Services - Instead of the Industry's Scope

Mindset: The 4th path takes its point of entry in going beyond maximizing the profit and value of products and services within one's own industry taking the company into a focus of finding complementary products and service to the customer in mind - finding and creating new complementary values to the customer.

Definition: The key in this 4th path is to define the full amount of values which the customers are looking for when they choose a product or a service. Therefore one should look at complementary products and services – and we would add – both before, under and after the buying process. The important questions to be asked are therefore: Within what sort of context are your products or services used by the customer? What happens before, during and after product and service use? Can you identify critical points related to why some buy or not buy your product? How can you eliminate these through complementary product or service offerings?

Case reflections: Existing products and services are at the center of the buying processes. Consumption cycle is at the center - before, during and after, ie. the consumption process is in focus and the values associated and complementary to this process are primarily addressed here. The "From cradle to grave" principle can also be transferred to this terminology. It is a parameter here to search for values which the customer may not have recognized yet or no one has demanded yet. The values sought here can in some context be similar to marketing literature terminology: "nice to have" - it may be that the client uses them in advance so we need to investigate the customer situation - similar to 3rd path.

Values addressed in this context are all values that are not already fulfilled by the company. Model of complementary products and service values:

Preliminary conclusion 4: The 4th path does not determine how far one may go with values – backwards and forwards in the consumption situation – consequently it requires a fairly comprehensive value analysis. The majority of companies in the research did not know or struggled to deal with the "before and after" situation - and its related values. They were often stuck with values related directly to their core business models.

*Figure 5: Dimensions addressing the customer's
buying situation of complementary values*

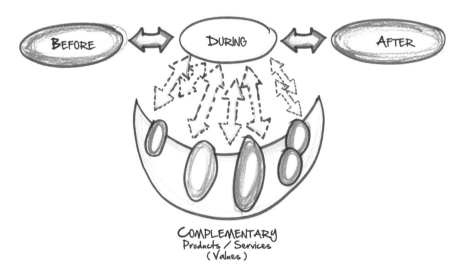

COMPLEMENTARY
Products / Services
(Values)

The 5th Path - The Functional and Emotional Orientation of an Industry – Instead of Same Orientation as Industry

Mindset: The focus of the 5th path is not just to focus on lowering the price within the same functional and emotional orientation of your industry but to find new value within physical and emotional area in your industry and hereunder possible "new objectives" to price.

Definition: Some industries compete on price and function only - primarily in utility calculations. Other industries compete primarily on emotion. In the Blue Ocean book the "Swatch" example is used where a whole industry changed from functional to emotional, or "QB" hairdressers in Japan, which focused on "high speed" haircut. The task is here to redefine the functional – emotional orientation of the industry and to find new values instead of a lower price. The first important question here to ask is: Are the industry competing on functional or emotional appeal? If you compete on emotional appeal, are there components you can remove to make it functional appealing? And oppositely; If you compete on functionality - what elements can you remove to make it emotionally appealing?

Case reflections: Moving the value set from "as-is" in to "to-be" corresponds to moving an entire community's focus from e.g. cost savings to value innovation.

The research at hand showed that this is indeed very difficult for a majority of companies – because it is perceived by the companies as related to high risk. The companies in our research had difficulties to perceive their company from another level and mindset and thereby change both their own attitudes and the attitude of the customers. In Denmark a commercial trying to get more danes to eat fish with the slogan "There are no bones in this" - "Gunnar and Minna" advertising spots for the fish industry) touch upon this challenge trying to play on other values in a mature industry.

Though the experience according to this path was that some companies are more agile to move their values from one context to another and at the same time incorporate both the emotional and the functional orientation. One carpenting company had values on social responsibility noticeable in their HR approach and was able to move these values into social innovation. This path in combination the other paths as e.g. the chain of customer and its context visualized new value and new service.

Values addressed: Basically this phase is connected to moving the focus of the client or clients in the industry from one set of value to another set of values.

Figure 6: Transforming from "as is" value set to "to be" value set

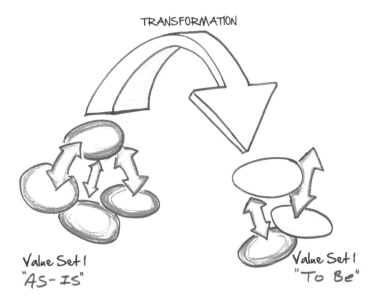

Preliminary conclusion 5: Truly addressing 5th path is maybe a trade mission/task concerning both physical and emotional values that can generate transformations of whole industries. However it is also a value set transformation which indeed can be very risky. Values are often interrelated and the company may not have the knowledge how they are interrelated.

The 6th Path - The Time Scope – Instead of Addressing the Same Time as Industry

Mindset: When competing - companies often focus on adapting to external trends. The 6th path's Blue Ocean focus is instead on future trends.

Definition: Whether you are a "first-mover" or a "follower" is an important question when you work with the 6th path. Think radically and not just incrementally. How do you think that the trends in the industry will change the future of your industry and your company? The advice is not to attempt to predict the future, but be aware of trends coming to influent your industry. Try to understand its effects. Try to look across time - from the value the market delivers today to the value the market delivers tomorrow. The Blue Ocean book mentions ITunes – Apple, who identified the potential of downloading digital music, when the rest of music business was busy preventing people from copying and sharing CDs.

The important questions to be asked are: What trends have a high probability of affecting ones industry. Is it irreversible and does proceed into a clear direction? How will these trends affect your industry? Given this - how can you open up new customer utilities?

Case reflections: In the case research which indeed represented SME companies, and not big companies as the book is exemplifying, this was a very big and challenging task to carry out. To support the companies to imagine future values and "may be" values turned out to be rather difficult.

Values addressed here are a combination of company core values in relations to emerging future values.

Preliminary conclusion 6: It was difficult to find new values that were meaningful to the companies – which our research showed is one fundamental condition if a company is meant to move from good ideas to action and implementation.

Figure 7: Transformation from "as is" value set to new and "may be" value set

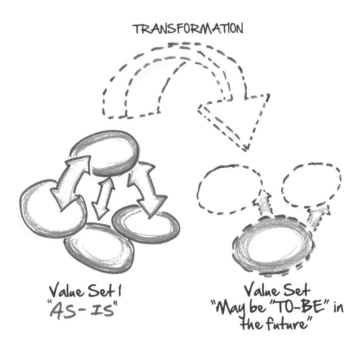

Case and Case Findings:
Immediate and consistent feature from the workshops showed that it was

* Difficult for many companies to see the relevance of all the Six Paths.
* Many companies found it difficult to relate to all Six Paths.

"We cannot get something out of this" where many companies' initial views. Specifically, the first path - alternative industries - was difficult for most companies to unfold in order for it to initiate new ideas.

The researchers found that in order to have full benefit from the paths one needed as innovation leader to promote a "translation" of the company and potential alternative industries. For instance related to the 1st path; a carpentering company is to be put within the building industry, but this gives not a full picture if you have to find alternative industries. It gives certain links to other industries as for instance do-it-yourself markets. But to encourage further ideas the researchers needed to be innovation leaders and interpret the company's values. The carpentering company had talked about their work with insurance companies. This gave the image of security; that one could construct and create more security in a chaotic world - eg. construction. They could look into their current customers for instance to check whether the work was well done, if extra repair were needed etc. and this way make an almost caretaking relationship to the customer. Further the company won a network price because they were very socially responsible towards its own employees. This also means that there are powers and hidden resources in the company which also could be used to "suit their customers and their values." One could consider this business to actually enter social innovative BMs. The researchers believed that this part of perceiving values from other perspectives was heavily downsized in the Blue Ocean Six Paths framework. One of the researchers stated that: "This path(1st) is not strong enough in the book according to really think alternatives. Even in the book one held onto ones own industry and the combination of the books reluctance to go out of the box does not inspire companies enough to do it".

Other interpretions were related to discover new profit formulas, new customer groups' values, the importance of intangible values vs. tangible values, service orientation vs. product orientation etc.

Discussion:

Framework in General

As the former example shows - interpreting the company's activities may be a way to use the tool better in order to redefine a specific market and develop the core business' values. But this also shows that Innovation leadership (Lindgren & Bohn 2003) is important for guiding the process fruitful. The researchers also saw this vs. interpreting the questions, so that the companies could relate the process to what was absolutely necessary in relation to all Six Paths. Continuous problems were quite similar vs. use of The Six Paths. The tool is good as an eye opener for the company and its leaders - but it needs interpretation and innovative leadership.

A few companies also thought the term "Six Paths" was misleading as a metaphor. They perceived this tool more as different aspects of the company and not paths to walk on.

Without innovation management firms still filled the questionnaire and participated in the discussions - but most of them (all?) were unable to challenge the use of the Six Paths so that it could initiate new markets and real new values.

6.2   Value in General

In order to use the value concept of the Six Paths framework program, the innovation leader needs to have a thorough understanding of the value framework, an "outside in" approach as well as an "inside out" approach to value innovation and understand the implications, challenges, barriers and pitfalls of defining values related to the Six Paths framework program. Further the innovation leader and the group involved in the value innovation process have to be able to support the creation of "meaningful value relations" (Saghaug & Lindgren 2009) throughout the process. To redefine values and find new BM for the company becomes only relevant as far as it is meaningful for both the potential customer and the company. Truly "thinking differently" (Lindgren and Saghaug 2008) from competitors addresses an emergent interpretation of the values related to both the existing BM and abilities to reach out into new potential values and new BM. Values addressed during the work with The Six Paths framework were defined in many different ways by the companies, and the evolving understanding by the companies of "values" as both related to tangibles and intangibles are not clearly supported in the Blue Ocean Book.

On the other hand our findings show that there are potentials in the Six Paths framework if the companies not fall into a "value trap", when working with it. This means, that if companies do not fully understand their values and how to thoroughly define (new) values related to the Six Paths framework program and align it to the potential value proposition of the market, customers and their company - they face a major strategic risk of losing – both resources invested in innovation and real possibilities to enter a new market.

Conclusion

The ability to understand the values and to commercialize these to real and new markets, real and new customer segments and the company opportunities related to these is important for the success of using the Six Paths framework program for the innovation of BMs and to improve the results of radical innovation successfully. However a misunderstanding of the Six Paths frame work, its value and value proposition can lead to serious strategic innovation faults.

What it means to innovate values according to the framework was reported under each path as the results were examined. Related to the different generic and perceived values of different alternative industries, markets, customers and companies, this offers a more detailed understanding than in other academic publications related to Blue Ocean tools. The values found influence the real and commercial opportunities of the blue oceans proposed. The table beneath gives an overview, sums up the findings and adds suggestions for the innovation leader.

Table 1: The Six Paths framework perspectives - an overview

The Six Paths framework perspectives	Values according to Blue Ocean Theory	Values according to findings in research	Pitfalls	Suggestions for innovation leaders	Cases
The 1st path - Alternative industries	Values are compatible across industries. Generalization	Values are not always compatible across industries. Diversification	Without interpretation" of values and innovation leadership this path is weak and not inspirational	Interpretations of "what are the company actually doing?"- working with –use of metaphors and visualizing new combinations across other industries	Sandmen, Handicrafts
The 2nd path - Strategic groups	Values may be moved from one position to another	Values are not obvious, but hidden. Values are convictions difficult to move.	Mindset - lack of meaningful relations to new value positions renders value positions unchanged	Challenge the mindset. Visualize opportunities	Harbor Lifters Sandmen

The Six Paths framework perspectives	Values according to Blue Ocean Theory	Values according to findings in research	Pitfalls	Suggestions for innovation leaders	Cases
The 3rd path – the Chain of customers	Values according to the chain of customer	Values are unknown and related to a different context	Lack of perspective: development of values over time. External influences on both part	Interpretation of values for different customer groups	Handicrafts Networker
The 4th path – complementary products and services	New values to the customer	Values only new to customer as far as values are a part of current value system in the company	Stuck with core business values.	As 3rd path and address customer situation. Develop understanding of company's core business values	Handicrafts
The 5th path - the functional and emotional orientation in industry	New value orientation	Values are not objects for change – moving value means transforming values	The idea of changing values is misleading unless they are a)transformed or b) part of current value set	Address potential hidden unknown values in the current value set or more radical - work on a transformation process-	Sandmen, Plumbers Harbor

The Six Paths framework perspectives	Values according to Blue Ocean Theory	Values according to findings in research	Pitfalls	Suggestions for innovation leaders	Cases
The 6th path – the time scope	Future trends that are clear and long lasting	Company core values in relations to emerging future values	Apart from major trends as energy savings etc. it was difficult to instrumentalise this path, unless it was meaningful in company's current value system	Use scenario techniques, Delphi techniques	Handicrafts Harbor

As shown above there were different views on values together with new implications, more challenges and pitfalls in connection to the use of values related to the Six Paths framework.

We verified that it is complicated to use the framework. Analysing the value chains, values and value proposition of the industries, markets and customers are very different, dynamic and also according to the companies, built upon a value framework that is often very different to their previous mindset. Also seen in "a time perspective" companies joining the research claimed that many values and opportunities will not turn up or will firstly be developed during the buying process. Further the industry/markets/customers value and value proposition of these are often not equal to the company's existing view on value and value proposition.

The findings indicates that a deep understanding of different values, value set, the values interrelatedness and success criteria related to different industries, markets, customers and companies also seen in a dynamic and time based view related to the innovation process will make a major influence and give a better access to the real potential of the proposed new BM. We believe that the above mentioned will significantly influence the realization of the outcome of the value innovation process – in the Blue Ocean six framework program. We expect to find that the ability to companies to act properly as a leader of innovation is related to that understanding of the Six Paths tools and the values and perceived values of importance for stakeholders involved in the innovation process.

This calls for a more precise and even new methods to encompass values of innovation projects related to the Six Paths framework program and value curve – and new tools for innovation leaders to apply the values in the Six Paths frame work program.

Perspectively further challenges are raised related to our understanding of economic metaphors (McCloskey 2001, 2009). This becomes evident as the understanding of "Value" in the Blue Ocean context and the Six Paths framework is rendered as a neutral concept. The missing elaboration of the value concept might be the biggest pitfalls of them all because this hinders a thorough understanding of the complexity of "values".

References

- Blois, K (2004), *Analyzing exchanges through the use of value equations The Journal of Business & Industrial Marketing;* 2004; 19, 4/5; ABI/INFORM Global

- Chan Kim, W. & Mauborgne, R. (2005), *Blue Ocean Strategy;* Harvard Business School Press.

- Chesbrough, H. and R. S. Rosenbloom (2000), *The Role of the Business Model in Capturing Value from Innovation: Evidence from XEROX Corporation's Technology Spinoff Companies',* Boston, Harvard Business School Press.

- Chesbrough, Henry (2006), *Open Business Models: How to Thrive in the New Innovation Landscape,* Boston: Harvard Business School Press.

- Christensen, C. M. (1997). *The Innovators Dilemma,* Harper Business ISBN 0-06-662069-4

- Christensen PR (2006) *På sporet af entreprenørskab og innovationsledelse - og det flagrende gardin, som adskiller de to rum,* in http://cjas.dk/index.php/loge/article/view/578/610

- Dougherthy, D., (1992) *Interpretative Barriers to Succesful product Innovation in large Firms,* Organisation Science. Vol 3, No.. 1992.

- IBM - Global CEO Stuy (2006), *Expanding the Innovation Horizon,* United Kingdom. http://www-935.ibm.com/services/uk/bcs/html/t_ceo.html

- Fals-Borda, O., & Rahman, M. A. (Eds.). (1991). *Action and Knowledge: Breaking the Monopoly with Participatory Action-Research (1st ed.),* New York: Apex Press.

- Gadamer, H.G. (1960) *Wahrheit und Methode (eng. Truth and Method),* Tübingen

- Lele, M (2005) *Monopoly Rules: How to Find, Capture, and Control the Most Lucrative Markets in Any Business* (Crown Books, 2005),

- Leifer, R. (2002) *Critical Factors Predicting Radial Innovation Success,* Berlin: Technische

- Lewin, K. (1946) *Action research and minority problems,* Journal of Social issues, vol 2,pp.34–46

- Lindgren., P., Saghaug K.F. & Clemmensen S. (2009) *The Pitfalls of the Blue Ocean Strategy - "The Importance of Value related to the Strategy Canvas",* CI Net 2009 Conference Paper.

- Lindgren., P , (2007) *"Innovating Business Models"* Cinet Gøteborg 2007

- Lindgren., P & Saghaug, K.F. (2008) *"Innovating business – a question of thinking differently".* Paper at Mopan .Boston, June-2008.

- Markides C. & C. Charitou. *"Competing dual business models: A contingency approach,* Academy of management executive, 2004. 18(3), pp 22–36

- McCloskey D. (2001) *Measurement And Meaning In Economics -The Essential Deirdre McCloskey,* Edvard Alger

- Porter, M. E. (1996), *"What Is Strategy?",* Harvard Business Review, vol. 74, no. 6, pp. 61–78.

- Saghaug, K.F. & Lindgren P. (2009) *"Implementing new strategies of operations in the intersections of academia and SMEs – with special focus on human beings as "differences" acting on relations towards meaningfulness",* Paper at the 16th International EurOMA Conference in Göteborg in June 09.

- Scharmer, C.O. (2001). *Self-transcending knowledge: sensing and organizing around emerging opportunities.* Journal of Knowledge Management, 5, (2) 137–151

- Scharmer, C.O. (2009) , *Theory U: Leading from the future as it emerges. The social technology of presencing,* Berret Koehler, San Francisco

- Tillich, P. (1951). *Systematic Theology Vol I, Reason and Revelation - Being and God,* Chicago, The University of Chicago Press.

- TRU Group at http://trugroup.com/whitepapers/TRU-Blue-Ocean-Strategy.pdf

- OECD (2007) *The Innovation Index.* OECD Annual report.

- Osterwalder, A. , Y. Pigneur and L.C. Tucci (2004), *Clarifying business models: Origins, present, and future of the concept,* Communications of AIS, No. 16, pp. 1–25.

- vSente,'s blog at http://twoscenarios.typepad.com/maneuver_marketing_commun/blue_ocean_strategy

- Wouters, M.J.F., Anderson, J.C. & Wynstra, J.Y.F. (2005). *The adoption of Total Cost of Ownership for Sourcing Decisions Elsevier*

- Wadsworth, Y *(1998) What is participatory action research?* Action research international is a refereed on-Action research is published under the aegis of the Institute of Workplace Research, Learning and Development, and Southern Cross University Press source: www.scu.edu.au/schools/gcm/ar/ari/p-ywadsworth98.html.

Chapter III.2

Network-Based Innovation – Combining Exploration and Exploitation?

René Chester Goduscheit

Introduction

An essential part of the debate on networks is the role of relationships and ties between the participating organisations. In his description of ties, Granovetter (1973) defines the strength of ties as a combination of the amount of time, the emotional intensity, the intimacy and the reciprocal services that characterise the tie. The remarkable result of Granovetter's analysis is that weak ties between individuals and between organisations are seen as indispensable to the actors' opportunities and to their integration into communities while strong ties between the actors lead to fragmentation of intergroup and interpersonal relations. From an inter-organisational network perspective, the work of Granovetter brings about a fundamental question: Should networks be built up by organisations that have already strong ties or relationships or should it be based on organisations with weak ties?

Granovetter focuses on positive, symmetric ties and, by this, excludes analyses of more pessimistic perspectives on ties and networks. The hopes of networks coexist with the fears of networks. While the networks, in some views, are the locus of positive things, such as innovation and learning (Jason & Powell 2004; Powell, Koput, & Smith-Doerr 1996; Tidd 1995), they potentially constitute loose, temporary and unreliable constructions, characterised by lack of long-term obligations (Waluszewski 2005). Thus, networks can be seen as a Janus head: The potential benefits of networking should be related to the potential costs and set-backs of engaging in networks. This balance of network considerations affects the perception of innovation in inter-organisational networks.

Intuitively, innovation in networks can be enhanced by two set-ups, which to a large extent are contradictory:

1. Networks created by organisations that do not know each other in advance and do not have insight into each others competences and resources are often breeding ground for innovativeness, complementary resources/competences and positive synergies. By bringing a variety of resources and competences together, one would expect the organisations to think 'out of the box'. Using Granovetter's terminology, these networks are built up by weak ties. Instead of focusing on the actors, with which an organisation has cooperated in the past and is currently cooperating, it should prepare a list of key parameters that are important in order to succeed in the future, identify organisations

that possess these resources/competences and contact these organisations (Preiss, Goldman, & Nagel 1996).

2. Networks composed by organisations that know each other very well from previous interactions and relationships and, by this, have deep insight into each other's resources and competences are able to utilise the innovation potential of each other. Such business relationships represent a variety of intangible resources that exist in the interface between the organisations such as skills, information, experience etc. (Gressetvold & Torvatn 2006). The cornerstone of the network is trust between the partners and a willingness to reveal innovative ideas within the network (Larson 1992).

The first set-up emphasises the plus-sum game of cooperating and networking and excludes considerations on opportunistic behaviour by the participating organisations. These considerations are immanent in the second set-up: The experiences of working together and the trust that results from these relationships reduce the (sense of) risk in innovating together.

This paper addresses some of the issues that arise when organisations innovate together in networks and attempt to balance the different considerations, which are outlined above. Thus, the research questions of the paper are:

- Is there a trade-off between the network innovativeness and the exploration of new possibilities, on one hand, and getting a specific outcome of the network at a given time frame, on the other?
- Do previous relationships between the network partners have an impact on this potential balance between innovativeness/exploration and outcome?

The two research questions address two different layers of inter-organisational networks. The first question is aimed at a discussion of the challenges of combining the two dimensions of networking: exploration and exploitation. The second question suggests a potential explanation of the difficulties in combining the two dimensions: Networking with organisations with which the focal organisation already has established relationships.

The paper is based on the KMD case in the NEWGIBM project. For a thorough description of the KMD case, please see elsewhere in this book.

The first part of the paper is a brief theoretical introduction to the key terms 'innovation' and 'networks'. The second part describes the overall methodological approach and operationalises the analytical framework. The third part is aimed at a discussion of the results of the case, which leads to the conclusion of the paper.

Innovation in Inter-Organisational Networks

The amount of literature on both innovation and inter-organisational networks is extensive. The connection between these two topics, innovation in inter-organisational settings, has also been analysed. The different perspectives from literature will be presented below and the paper will be related to the literature.

Interactions, Relationships and Networks

The term 'network' is often very loosely used to describe any relation between individuals, groups, organisations etc. Thus, the broad understanding of networks covers everything from an executive's 'black box' of useful contacts to an integrated company organised on internal market lines (Child, Faulkner, & Tallman 2005). There are, however, three central components of a network: actors, activities and resources (Haakansson 1990; Haakansson & Snehota 1989). These three aspects influence and control each other in a complex interplay. The actors control resources, the actors perform activities, and the activities link the resources together.

The literature on inter-organisational networks differs in the view of the ties between the participating organisations. Ford et al. (1998) analyse the relations between organisations as consisting of different bricks. On the lowest level, one finds interactions between the organisations. Such interactions are the exchange of products, services, money or social 'chit-chat'. In other words, interaction includes both interpersonal communication and interaction through the delivery of physical products and services, information and payments (Ford & Haakansson 2005). These interactions are regarded as episodes in the total relationship between the organisations, which is the next level of relation between organisations. Thus, the relationship between two organisations consists of the previous episodes and the effect of these episodes on the future ones. The relationships of one organisation are tied up in a complex, interdependent network with a larger number of organisations – the broadest level of relation between organisations (Ritter & Gemünden 2003a). Each of these organisations have their 'own' network, which make the network structure even more complex.

Other researchers have contrasting views to this build-up approach to networks (Braha & Bar-Yam 2005; Powell, Koput, & Smith-Doerr 1996;Preiss, Goldman, & Nagel 1996; Sydow & Windeler 1998). They focus on more fluid, transitory and informal aspects of inter-organisational activities as the fundaments of a network. Thus, they do not presuppose the existence of relationships between organisations and/or actors in their definition of network. Some researchers even speak of the spontaneous formation of a network, which is marked by a loose and changeable coupling of the participating organisations (Konsinsky & Amrit 2005).

Ford (1990) describe the development of buyer-seller relationships in terms of uncertainty, distance (socially, geographically-culturally, technologically and time) and commitment. The early stages are defined as the period where potential suppliers are in contact with the purchasers to negotiate or develop specifications for a given good. These stages are marked by high uncertainty, high distance and low commitment. In the long-term and final stages, both the uncertainty and distance are reduced while the commitment to the relationship is increased. Though Ford's description is based on the buyer-seller relationship, the terminology can be applied to other sorts of networks, which will be demonstrated in the empirical part of the paper.

Innovation

One of the things for an organisation to consider when initiating an innovation process is the ambitions of the process. The organisation might want to set up a process that searches for new possibilities, risk taking, experimentation, flexibility in the process, discovery of new, potential business plans and innovation. Alternatively, the organisation might wish to set-up a process, that is aimed at getting the most out of the existing resources in the organisation and refiningthe current procedures. March (1991) has termed this distinction exploration versus exploitation. Exploration projects are generally marked by much higher uncertainty (in terms of financial returns, time-frame for obtaining a result etc.) than exploitation projects. This makes the exploration projects more vulnerable than the exploitation projects.

Literature on innovation has made similar observations as to the distinction exploration and exploitation. Dewar and Dutton (1986) describe the difference between radical and incremental innovation. Radical innovations are fundamental changes that represent revolutionary changes in technology and a clear departure from existing practices. In other words, radical innovations embody a new technology that results in new market infrastructure. They often do not address a recognised demand but, instead, create a demand previously unrecognised by the customer (Garcia & Calantone 2002). In contrast, incremental innovations are minor improvements or simple adjustments in current technology. These innovations provide new features, benefits, or improvements to the existing technology in the existing market. The distinction between radical and incremental innovations is by nature somewhat arbitrary (It is difficult to give a conclusive definition of new/existing technology/markets) and literature reviews have identified discrepancies in making clear-cut definitions of the character of the innovation. However, if the radical/incremental variable is seen on a continuum scale instead of as on a dichotomous scale, the variable can be useful in understanding different kinds of innovation.

A third distinction in the literature applies to the goal of the innovation process. Trist (1983) distinguishes between the goal as a 'problématique' and the goal as a discrete problem. While the problématique is concerned with a rather overall problem (environmental, societal etc.), the discrete problem is aimed at a clearly defined problem, that needs a solution (and often within a short time frame).

Networks and Innovation

A network with many participants generally contains substantial innovation resources (by virtue of the resources of each individual participant). Inter-organisational networks are generally acknowledged as the locus for innovation and learning. Thus, when the knowledge base of an industry is both complex and expanding and the sources of expertise are widely dispersed, the innovation and learning can be derived from inter-organisational networks rather than within individual organisations (Powell, Koput, & Smith-Doerr 1996).

The coordination and cooperation of numerous actors, however, is a complex game, in which issues like power, distribution of benefits and costs, intellectual property rights, politics etc. are innate e.g. (Cook & Emerson 1978; Swan & Scarbrough 2005). These issues can hinder the exploitation of the resources within the network and the (willingness to carry out) exploration of new possibilities. A lesser network that does not have the same innovation resources can, on the other hand, be more manageable for a focal organisation, thus making it more easy to get a relatively rapid outcome from. Gulati & Gargiulo (1999) describe the importance of relational embeddedness. This term stresses that prior experience between two or more organisations has a major importance for future cooperation: The probability of a new alliance between two organisations increases with the number of prior direct or indirect alliances between those organisations.

This chapter regards the interface between innovation and networks. In order to understand the development of the analysed network, both innovation and network literature will be included, but the prime concern is the interface between the two traditions.

Methodology

This chapter is based on a single case network, that was chosen on the basis of the potential theoretical contribution of the findings in the case (Eisenhardt 1989). Thus, the KMD network was selected due to two interesting aspects of the network. Firstly, the network consists of both partners who are completely new to each other and partners between whom relationships were already established at the outset of the network. To some extent, two networks have coexisted within the same project (see the description of the core and peripheral group in the description of the KMD network elsewhere in the book). This opens up the possibility of making comparisons of the dynamics of the two networks. Secondly, throughout the entire process, the focal organisation has been very committed to developing their platform in the network and letting the researchers be an integrated part of this process. Thus, the likelihood of finding interesting perspectives in relation to the two research questions seemed considerable.

Four network meetings have been selected to describe the case. There have been other meetings within the network: The research group met with the focal organisation numerous times and the focal organisation has had bilateral meetings with the core group members (as a part of their general business relationship – see description below). All of these meetings could potentially have had an impact on the development of the network. Four particular meetings have been selected because they represent crucial episodes in the network development. Furthermore, the four meetings illustrate the research questions addressed in this paper: The character of the innovation within the network and the (potential) impact of previous relationships in the innovation process.

Data

The author of this chapter and four other researchers involved in the project have been actively involved in the Intelligent Utility Network as participants at network meetings and as sparring partners in the network formation and development. At the network meetings, several researchers participated and the observations were triangulated after the meetings.

In addition to accumulating observations, the author, together with a research colleague, carried out numerous interviews with the different organisations involved in the network. These in-depth interviews were recorded and transcribed and the interpretation of the interviews were triangulated in the researcher group as well.

Operationalisation

As mentioned above, the four meetings can be regarded as critical events in the network development and as an indication of how the network has been working. With the view to systemise the presentation of the network development and to focus on the theoretical aspects in question, four parameters are introduced:

Output: the number of ideas which have been generated at the meetings. This is merely a quantitative measure and does not look at the character or quality of the generated ideas.

Exploration rate: the degree to which the generated ideas are explorative. In recognition of the arbitrary character of this variable, the original vision paper of the director of the focal organisation is used as a benchmark: If the ideas generated at the meeting are beyond the visions of the paper, the exploration rate is high. If the idea generation is more limited than the initial paper, the score is low.

Outcome: the level of commitment from the participants in terms of progressing with the generated ideas. This variable indicates the ability of the participating organisations to provide the network with resources to pilot the developed solution (end-users, technology, manpower, direct funding etc.). In addition, it encompasses the commitment of the partners in terms of applying for external (EU or national) funding. The term outcome is by in large equivalent to Ford's (1990) term 'commitment'. However, the variable is not merely aimed at the level of commitment to the other partners but also to the innovations (services, goods, concepts, etc.) of the network.

Coherence: the level of coherence between the participating organisations in terms of previous relationships. If the network meeting is based on participants, who all know each other from previous relationships, the coherence is estimated as high. If the network meeting participants are all new to each other in terms of previous episodes or relationships, the coherence is estimated as low. The number of participants has an impact on the coherence as well. The coherence of a network consisting of a relatively limited number of participants tends to be higher than networks with a larger number of participants. The term coherence has similarities to Ford's (1990) term distance. The coherence variable, however, is beyond the dyadic relationships and is aimed at the entire network and its density – not merely the social, geographical/cultural, technology and time distance between two partners.

All the four variables are indexed in a continuum between 0 (low) and 1 (high). The scoring is subjective because of the nature of the four variables. If 1 is the maximum score on output, the equivalent number of ideas for scoring 1 on that scale is difficult to determine. It could be 10, 20 or even 200 ideas at a meeting. The scoring has been carried out, however, partly on the basis of the interviewees' assessment of the output, outcome, exploration and coherence of the particular meetings and partly on the basis of the researchers' experiences from similar inter-organisational networks. For instance: Which exploration of new concepts, services and products have we seen in other networks, and what is the degree of exploration at this meeting?

Figure 1 illustrates the network development in terms of the four variables throughout the network development from the initial meeting in March 2006, to the last meeting in February 2007. The next chapter will discuss the implications of the network development.

Analysis

The analysis of the Intelligent Utility Network is divided into a discussion of the overall development of the network, a description of the content of the network innovation, a discussion of the impact of previous relationships in the network, and concluding remarks on the success of the network.

Figure 1: Overview of Network Development
from March 2006 - February 2007

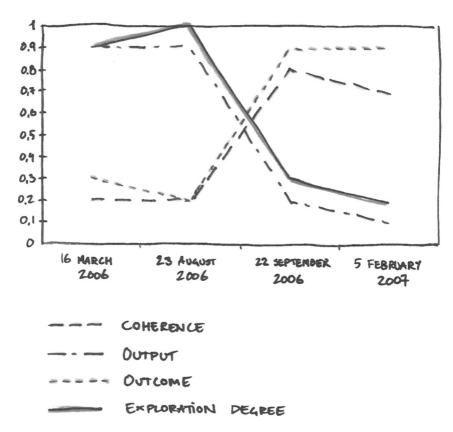

Discussion of the Overall Development of the Network

Figure 1 illustrates the development of the network based on experiences from the network meetings and interviews with the participating organisations.

The March 2006 meeting was characterised by a large number of participants, who to a large extent were brought together without knowing each other beforehand. While the coherence of the participants was limited, the output in terms of ideas for the Intelligent Utility Network was very high. The content of the ideas was explorative compared to the original vision paper of the focal organisation, bringing new and innovative dimensions to the platform (services within social care for elderly people).

Principally, the August 2006 meeting can be described as similar to the March 2006 meeting. Some of the generated ideas, however, were even more explorative than the ideas at the previous meetings (visions on the functionalities within the social service area and purchasing organic energy were prepared and included in the envisaged future platform). On the other hand, the outcome was limited. None of the participating organisations expressed their interest in actually committing themselves to taking part in the funding of developing the platform.

As described above, the set-up of the September 2006 meeting differed from the previous two meetings. Three utility providers, with which the focal organisation already had established relationships, were invited. The degree of outcome and exploration degree had declined compared to the first two meetings: The utility providers were, to a large extent, 'thinking in the box' of the current need of the customers. Due to the fact that the focal organisation had existing relationships with the utility providers (and that the utility providers knew each other from professional networks in the utility industry), the coherence of the narrow network was high. The outcome of the meeting was high as well: The focal organisation and the utility providers decided on a time schedule for setting up the platform and agreed on a market introduction deadline.

The February 2007 meeting was arranged as a workshop, which would clarify the exact functionalities and modules of the Intelligent Utility platform. The participants were the same as in September 2006. However, two additional utility providers were invited to the meeting. The focal organisation had already established a relationship with one of these 'new' utility providers but the other utility providers did not know the representatives of the organisation. The other 'new' utility provider had no previous relationship with the focal organisation. Thus, the coherence of the participating organisations was lower as compared to the previous meeting. The outcome of the meeting was not different from the September 2006 meeting: The same three utility providers were still holding on to the former time schedule and market introduction deadline. The two 'new' participating organisations were not invited to participate in implementing the platform. The degree of output and exploration was decreasing as in relation to the September 2006: The smart home functionalities were ruled out by both the focal organisation and some of the core partners as being too premature for the platform.

Innovation in the Intelligent Utility Network

As described in the previous section on the case, the focal organisation, at an early stage, recognised that they had to involve other organisations in order to realise the visions of the Intelligent Utility project. However, the different actors in the focal organisations didnot seem to agree fully about the intentions behind and implications of setting up the network. The area director, who was preparing the vision paper behind Intelligent Utility, perceived the network as a way to collect new ideas and visions for the final service platform. Thus, the explicit commission of the network was to generate new and visionary ideas. The perception of the project leader in the focal organisation differed from this 'idea factory' understanding of the network. He primarily referred to the network as a way to implement the visions of Intelligent Utility, which is condensed in the following statement:

"We knew that all the technical aspects of our vision [as stated in the vision paper] already existed. We just needed to find partners that would provide us with the technology and the end-users for the product..... The project was merely an implementation and not a development project".

The discrepancy of perception of the function of the network influenced the work-flow and development between and during the network meetings. While the director constantly tried to open up the network to encompass organisations that could add new angles and visions to the service platform, the project leader was keen on ensuring outcome and moving to prototyping. As a consequence of the internal discrepancy in the focal organisation, the Intelligent Utility project turned out to be based on, not one, but two separate and dissimilar networks. Figure 2 illustrates the two networks and the dissimilarities.

The meetings of the comprehensive network, which consists of the core and peripheral groups in the network, were marked by a focus on idea generation, exploration of new possibilities and radical innovations. There were two meetings in the comprehensive network: one in March 2006 and the other in August 2006. The August 2006 meeting generated a long list of potential (and currently technically feasible) solutions on the Intelligent Utility platform: Self-sufficiency in energy consumption (the private home as a micro energy plant); purchasing of organic energy; and social services for certain customer segments (care for elderly citizens and disabled people). In general, the meetings in the comprehensive network were marked by a focus on societal 'problématiques' and general sustainability, such

Figure 2: Overview of Characteristics for the
Comprehensive and the Narrow Network

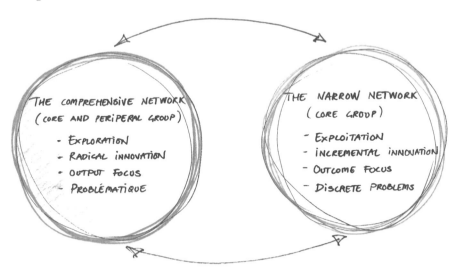

as lowering the release of carbon dioxide, handling future challenges of elderly people and maximising the sense of security in society.

The meetings in the core network were by all means different from the ones in the comprehensive network. These meetings, in September 2006 and in February 2007, were marked by strict agendas, focus on concrete outcomes from the meetings ('What do we do now to accomplish what we've decided?') and handling of discrete problems ('Which customer need do we want to accommodate and is the customer willing to pay for it?').

One could expect some degree of spillover from the comprehensive network to the core network: Some of the ideas generated in the explorative processes in the comprehensive network could potentially inspire the processes of the core group. Thus, representatives of the core group organisations participated in both of the comprehensive network meetings. Apparently, the effect has been the opposite. The ideas and visions about the integration of smart home solutions into the platform were neglected and, to some extent, labelled as 'a pie in the sky'. At the September 2006 and the February 2007 meetings, the project leader of the focal organisation agreed with the other participants to disregard the smart home solutions because of the 'current irrelevance'.

Previous Relationships in the Network

The Intelligent Utility Network has been marked by openness throughout all of the phases. At an early stage, the focal organisation decided that all interested persons and organisations could attend the network meetings if they could state the case for their interest and potential contribution to the network.

The peripheral organisations in the network were involved in various ways. The focal organisation had previous relationships with two of the peripheral organisations and contacted them in order to involve them in the Intelligent Utility Network. Two peripheral organisations were a part of the network of the moderator/organiser of the August 2006 meeting and were involved in the network through this connection. The last three organisations were involved through the initial contact via mails and letters from the focal organisation and the researchers in January 2006.

The narrow network (the core group) was marked by strong relationships and coherence between the focal organisations and the other organisations involved. Though the focal organisation had previous relationships with a few of the peripheral organisations, the comprehensive network was a much looser network composition in terms of relationships.

The meetings in the core group illustrated how previous and ongoing relationships, to a certain extent, can be an impediment to explorative, innovative processes. The utility providers were constantly relating the new platform to the existing services of the focal organisation. The following debate of the September 2006 meeting demonstrates this point. The project leader of the focal organisation made a thorough presentation of the overall perspectives of the envisaged platform. The presentation was aimed at the problématique, that the solution was trying to address (the societal benefit from lowering the energy consumption of the private end-users). As mentioned above, the utility providers in the core group were already using the data processing system of the focal organisation to invoice the private customers and were focusing on the potential overlap between the new solution and the existing service, for which they were already paying the focal organisation:

> 'Does this mean that we are going to pay twice? [...] We are already receiving this service from you'.

In addition, the meetings in the core group tended to be rather operational. The utility providers constantly asked about the data structure and system architecture. The specialists of the focal organisation with thorough insight into the technical side of the platform tended to be predominant compared to the project leader, who was a generalist. This meant that the overall, strategic perspectives, to some exten, were excluded from the agenda, while the discrete, operational problems of the existing services and the future platform took over.

In spite of the fact that the participants in the core group were keen on focusing on the details and the discrete problems of the solution, the group worked very efficiently and outcome oriented. It was not an issue of whether to cooperate but rather how to cooperate.

Has the Network Been a Success?

The pending issue is whether or not the Intelligent Utility Network has been a success. Since the primary aim of the network (at least in the area director's opinion) was to utilise the network to innovate, one angle could be to assess the innovation success of the network. However, the discussion of network success is just as complex as the discussion of the different aspects of innovation (radical/incremental, exploration/exploitation etc.). Some analyses have operationalised innovation success as rating the performance of the firm compared to the competitors and the 'state of the art' (Ritter & Gemünden 2003b; Ritter, Wilkinson, & Johnston 2002). In these analyses, the rating is based on the self-assessments of the firms.

The Intelligent Utility Network is a success in terms of getting an outcome from the network. The focal organisation and the restricted network are in the process of developing a prototype of the platform and they are trying to collect external funding for the project. All the participating organisations are trying to make the most of their individual assets through new ways of combining these assets: The focal organisation has a large amount of data on potential customers and methods of processing this data. The utility providers have direct contact to a large number of customers. In other words, by combining their individual assets, the network partners are seeking (and apparently succeeding) to exploit their existing competences and resources. The impression given from the meetings and the interviews with the partners is that the organisations assess themselves and the Intelligent Utility Network as being very innovative as compared to the 'state of the art' and other competitors.

The innovation success of the network, however, can be questioned if one in-cludes parameters beyond exploitation of the resources and self-assessment of the participating organisations. As becomes clear in the prior discussion of the network development, only a limited number of companies out of the compre-hensive network are actually involved in prototyping the platform. Compared to the original vision paper of the area director of the focal organisation and the idea generation of the comprehensive network, the explorative innovation in the final prototype is limited. The prototype cannot be regarded as a radical in-novation since the solution does not include new technology that results in new market infrastructure. As mentioned above, the incremental/radical dimension should not be seen as a dichotomous variable but rather a continuum. However, concerning the novelty of the prototype for the market and in terms of technol-ogy, the platform seems to be closer to the incremental than the radical end of the continuum.

An assessment of the network success should, thus, clarify on which parameters the assessment is based. If an optimal exploitation of the existing competences and resources in the core network is the parameter, the network is a success. Alternatively, the assessment the network success can be based on the counter factual question: 'What could have been achieved if the focal organisation man-aged to include the comprehensive network and utilise all or some of the ideas generated in the March 2006 and August 2006 meetings?' From this perspective, which emphasises the aspects of exploration and radical innovations, the network success is more dubious.

Conclusion

The chapter attempts to make two contributions. The first is an illustration of the challenges from a network perspective to balance the exploration of new possibilities, on one side, and actually getting a concrete outcome of the network within a given time frame. The second is to enhance the understanding of the effects of basing the inter-organisational network on partners that already have established relationships.

In relation to the first contribution, the Intelligent Utility case illuminates con-siderable challenges in combining innovation and exploration with being able to move forward and getting a concrete outcome from the networking. The high ambitions of the focal organisation, in terms of exploring new possibilities, to-gether with the network partners contrast with the actual result of the network.

Though the 'scaling down' of the explorative side of the platform could be seen as a deliberate decision of the project leader, the case seems to indicate that network partners find it challenging to explore new possibilities. To some extent, the project leader makes a virtue of necessity when lowering the explorative ambitions. The comprehensive network cannot combine exploration with actually getting an outcome of the network. The exploration of the core group is impeded by the perception that the provided services should be feasible, tangible and demanded by the customers.

The Intelligent Utility case demonstrates a parallel network structure in the comprehensive network and the core group. Though the organisations of the core group participate in the comprehensive network as well, there is no spillover from the exploration and idea generation of the comprehensive network to the core group. Actually, the ideas of the comprehensive network are labelled as a 'pie in the sky' and are not considered as relevant to the platform.

The second contribution is an indication of how previous and ongoing relationships between partners influence the inter-organisational network. The case suggests that the efficiency of a network consisting of organisations that know each other from established relationships is considerably higher than networks which are based on 'new' organisations. The core group of the Intelligent Utility Network is marked by a high degree of outcome in terms of commitment from the organisations to finance the platform development and trying to get external funding. The previous and ongoing relationships in the core group, however, tend to hamper idea generation and the exploration of new possibilities within the provided service.

It is clear that the two contributions of the paper are by no means isolated from each other. The Intelligent Utility Network illustrates that in order to ensure the progress of the network in terms of getting a specific outcome, the focal organisation has a tendency of choosing the 'easy', incremental innovation. Thus, through the already established relationships in the core group, the focal organisation develops tangible and specific services, and the focal organisation neglects the possibility of making explorative, radical innovation in the comprehensive network.

It should be stressed that this paper is not an attempt to grade 'innovation quality'. Incremental innovation and exploitation is not ranked as inferior to radical innovation and exploration. The platform developed by the core group is most likely to be innovative and add new value to the customers. However, compared to the initial visions of the Intelligent Utility platform, the solution developed will not be fulfilling the expectations. In other words, inter-organisational networks can be the locus of innovation. But the nature of the innovation of the network depends on a long list of aspects, such as characteristics of the participating organisations, the project leadership and the time frame for the final outcome of the network process.

Acknowledgement

The author would like to thank the focal organisation and the other participating organisations for their willingness to involve the researchers in all stages of the network. The study is a part of the NEWGIBM project, which is co-funded by the Danish Ministry of Science, Technology and Innovation and the participating organisations. The author is grateful for the comments and suggestions of the NEWGIBM research group in preparing this paper.

References

• Braha, D. & Bar-Yam, Y. 2005, 'Information Flow Structure in Large-Scale Product Development Organizational Networks,' in *Smart Business Networks*, P. Vervest et al., eds., Springer, Berlin, pp. 105–125.

• Child, J., Faulkner, D., & Tallman, S. B. 2005, *Cooperative Strategy*, Oxford University Press.

• Cook, K. S. & Emerson, R. M. 1978, "Power, equity and commitment in exchange networks", *American Sociological Review*, vol. 43, pp. 721–739.

• Dewar, R. D. & Dutton, J. E. 1986, "The Adoption of Radical and Incremental Innovations: An Empirical Analysis", *Management Science*, vol. 32, no. 11, pp. 1422–1433.

• Eisenhardt, K. M. 1989, "Building Theories from Case Study Research", *Academy of Management Review*, vol. 14, no. 4, pp. 532–550.

• Ford, D. 1990, "The Development of Buyer-Seller Relationships in Industrial Markets," in *Understanding Business Markets: Interaction, Relationships and Networks*, D. Ford, ed., Academic Press, London, pp. 42–57.

• Ford, D., Gadde, L.-E., Haakansson, H., Lundgren, A., Snehota, I., Turnbull, P., & Wilson, D. 1998, *Managing Business Relationships* John Wiley and Sons Ltd, Colchester.

• Ford, D. & Haakansson, H. 2005, "The Idea of Business Interaction", *IMP Journal*, vol. 1, no. 1, pp. 4–27.

• Garcia, R. & Calantone, R. 2002, "A critical look at technological innovation typology and innovativeness terminology: a literature review", *Journal of Product Innovation Management*, vol. 19, no. 2, pp. 110–132.

• Granovetter, M. S. 1973, "The Strength of Weak Ties", *The American Journal of Sociology*, vol. 78, no. 6, pp. 1360–1380.

• Gressetvold, E. & Torvatn, T. 2006, "Effects of Product Development: A Network Approach", *IMP Journal*, vol. 1, no. 2, pp. 60–84.

• Gulati, R. & Gargiulo, M. 1999, "Where Do Interorganizational Networks Come From?", *The American Journal of Sociology*, vol. 104, no. 5, pp. 1439–1493.

• Haakansson, H. 1990, "Technological Collaboration in Industrial Networks", *Engineering Management Journal*, vol. 8, no. 3, pp. 371–379.

• Haakansson, H. & Snehota, I. 1989, "No business is an island: the network concept of business strategy", *Scandinavian Journal of Management*, vol. 5, no. 3, pp. 187–200.

• Jason, O. & Powell, W. W. 2004, "Knowledge Networks as Channels and Conduits: The Effects of Spillovers in the Boston Biotechnology Community", *Organization Science*, vol. 15, no. 1, p. 5.

• Konsinsky, B. & Amrit, T. 2005, "Spontaneous Collaborative Networks," in *Smart Business Networks*, P. Vervest et al., eds., Springer, Berlin, pp. 75–89.

• Larson, A. 1992, "Network Dyads in Entrepreneurial Settings: A Study of the Governance of Exchange Relationships", *Administrative Science Quarterly*, vol. 37, no. 1, pp. 76–104.

• March, J. G. 1991, "Exploration and Exploitation in Organizational Learning", *Organization Science*, vol. 2, no. 1, pp. 71–87.

- Powell, W. W., Koput, K. W., & Smith-Doerr, L. 1996, "Interorganizational Collaboration and the Locus of Innovation: Networks of Learning in Biotechnology", *Administrative Science Quarterly*, vol. 41, no. 1, pp. 116–145.

- Preiss, K., Goldman, S. L., & Nagel, R. N. 1996, *Cooperate to Compete. Building Agile Business Relationships*, Van Nostrand Reinhold, New York, NY.

- Ritter, T. & Gemünden, H. G. 2003a, "Interorganizational relationships and networks: An overview", *Journal of Business Research*, vol. 56, no. 9, pp. 691–697.

- Ritter, T. & Gemünden, H. G. 2003b, "Network competence: Its impact on innovation success and its antecedents", *Journal of Business Research*, vol. 56, no. 9, pp. 745–755.

- Ritter, T., Wilkinson, I. F., & Johnston, W. J. 2002, "Measuring network competence: Some international evidence", *Journal of Business and Industrial Marketing*, vol. 17, no. 2–3, pp. 119–138.

- Swan, J. & Scarbrough, H. 2005, "The politics of networked innovation", *Human Relations*, vol. 58, no. 7, p. 913.

- Sydow, J. & Windeler, A. 1998, "Organizing and Evaluating Interfirm Networks: A Structurationist Perspective on Network Processes and Effectiveness", *Organization Science*, vol. 9, no. 3, pp. 265–284.

- Tidd, J. 1995, "Development of novel products through intraorganizational and interorganizational networks the case of home automation", *Journal of Product Innovation Management*, vol. 12, no. 4, pp. 307–322.

- Trist, E. 1983, "Referent Organizations and the Development of Inter-Organizational Domains", *Human Relations*, vol. 36, no. 3, pp. 269–284.

- Waluszewski, A. 2005, "Hoping for Network Effects or Fearing Network Effects", *IMP Journal*, vol. 1, no. 1, pp. 98–118.

Chapter III.3

Innovating New Business Models in Inter-firm Collaboration
– A Closer Look on the Fuzzy Front-End

Jacob Høj Jørgensen
René Chester Goduscheit
Carsten Bergenholtz
&
Peter Lindgren

Introduction

Innovation and innovation processes have traditionally been considered from the manufacturing company's perspective. Whether it is product-innovation, process-innovation or the innovation of a new business model, the innovation process is typically divided into a series of succeeding stages where the 'Fuzzy Front-End' is the first stage to confront. Several research projects have formulated recommendations for the company to improve the innovation process and enhance the chances of success. The vast majority of these projects, however, belong to an intra-firm paradigm where the focal company is considered to be the only part involved in the process, controlling and influencing the environment (Cooper 2005; Cooper & Kleinschmidt 1987; Tidd, Bessant, & Pavitt 2005).

As a result of enhanced competition and pressure on prices, increasing focus is put on inter-firm collaboration and innovation competences. Companies can engage in such inter-firm collaborations in regard to many different activities e.g. logistics, marketing and sales. The focus of this chapter is on inter-firm collaboration where innovation of a new business model is the main part of the collaborative effort.

Formal innovation partnerships have been widely researched. (Bart Nooteboom 2003; Faems, Van Looy, & Debackere 2005; Hagedoorn 2002; Powell, Koput, & Smith-Doerr 1996). The research has provided useful insights into the dynamics and tendencies in formal R&D partnering relations. This chapter, however, focuses on collaboration between independent companies prior to such formal agreements as joint ventures or other contractual agreements.

This first phase of the innovation process is often referred to as the 'Fuzzy Front-End' (FFE) and is traditionally seen as an intra-firm process (Jongbae & David 2002; Kim & Wilemon 2002; Qingyu & William 2001; Reid & de Brentani 2004). As the innovation process becomes an inter-firm collaboration, the management of the FFE also changes and calls for new ways of collaboration. This chapter examines the characteristics of the FFE phase and explores this phase in an inter-firm perspective. Through an in-depth case-study of a single firm and its innovation partners, parameters for improved collaboration in the FFE are identified. In this case, the new business model is highly reliant on the partner's ability to innovate a new product and thus focus is on this process.

The objective of this chapter is to elaborate on the differentiating characteristics between intra-firm and inter-firm FFE projects. Focus is on the management methods of the collaboration and the CEO-commitment to the project.

Firstly, the methodological approach is described. Secondly, a discussion of the collaboration dichotomy is carried out. Thirdly, the FFE phase is characterised through a literature review followed by a brief case description. Finally, the case is analysed in relation to the factors *management methods, formalisation* and *CEO-commitment*. This is done in order to reveal differences in going through the FFE in an intra-firm and an inter-firm setting.

Methodology

The chapter will be based on a case study of a Danish inter-firm network within the energy sector. The focal firm KMD, a major Danish IT provider, wished to enhance its penetration in the B2C market by providing internet services on energy consumption. In order to accomplish this vision, the focal firm initiated a network of firms potentially in a situation to contribute to a new business model. During a period of one year four network meetings were carried out. A maximum of 12 firms participated but, on some occasions, only a limited number of firms took part.

In order to get an in-depth understanding of the dynamics involved in an inter-firm FFE project, the case study method has been applied, in accordance with the guidelines set forth in (Eisenhardt 1989; Eisenhardt & Graebner 2007; Flyvbjerg 2006; Yin 1994).

The authors have been actively involved in the network as participants, sparring partners and observers at network meetings in the network formation and development. To increase the validity of the research, different sources of data have been triangulated: Documents, observation and interviews. The observations have also been triangulated, since different researchers with differing theoretical standpoints have been present at the network meetings. Subsequent to all meetings, the observations and reflections were discussed. At the end of the FFE phase, the involved networking partners participated in in-depth interviews. The questions in these interviews have been related to theoretical constructs, and not been aimed at verifying or falsifying specific relations between parameters. All interviews have been recorded and transcribed, and interpretations of the interviews have been discussed by at least three researchers. The aim of these triangulating actions has been to ensure that all relevant alternative interpretations have been included.

Overall, the case study method has been used to describe the relevant parameters for inter-firm collaboration in FFE projects, and, furthermore, it has been explained why these parameters are relevant.

Collaboration

Powell (1990) presents a taxonomy of three overall forms of collaboration: Hierarchy, networks and markets.

Table 1: Three Forms of Collaboration (Powell 1990)

	Hierarchy/ Intra-firm Collaboration	Network/ Inter-firm Collaboration	Market
Normative Basis	Employment Relationship	Complementary Strengths	Contract, Property Rights
Means of Communication	Routines	Relational	Prices
Tone or Climate	Formal/Bureaucratic	Open-ended, Mutual benefits	Precision and/or Suspicion
Actor Preferences or Choices	Dependent	Interdependent	Independent
Methods of Conflict Resolution	Fiat/Supervision	Reciprocity and Reputation	Haggling

Table 1 clarifies the essential differences between the three kinds of collaboration. Hierarchical and market collaboration are located at opposite ends of the continuum while networks are a hybrid between the two extremes. The distinction between hierarchy and network is equivalent to the difference between intra-firm and inter-firm relations. Though some hierarchy might exist between two firms in a network (in terms of size, intellectual properties, economic and staff resources etc.), the normative basis, communication, tone etc. will differ from the intra-firm collaboration. Thereby, the managerial implications of handling the FFE also differ from an intra- to an inter-firm perspective.

The inter-firm network, for instance, is not founded on the same degree of routines and formal tone as the hierarchy form of collaboration – and direct means of power, such as fiat and supervision, will not be feasible in a network set-up. The reciprocity, interdependence and complementary relationship between the firms involved in the network make the sources of influence and power much more subtle.

What is The Fuzzy Front-End?

The Fuzzy Front-End (FFE) is the first phase of the innovation process and initiates the process by producing ideas for incremental or radical product or service concepts. The term 'Fuzzy' refers to the intangible nature of this particular stage of the innovation process. It is considered fuzzy for a number of reasons. Examples are uncertainties and unknown issues concerning the needs of the customers, uncertainty about what competitors are doing, and uncertainty about which product and process and/or technologies should be used. Uncertainty concerning strategy alignment, internally and between partners, required resources, capabilities and company limits also prevent an opportunity from going on to the more structured New Product Development (NPD) phase (Kim & Wilemon 2002).

Thus, many of the practices that are used in the NPD phase do not apply to the FFE. They fall short because the nature of work, activities, funding level, revenue expectations, and measures of progress are fundamentally different (Koen et al. 2002).

The FFE is of interest because it has a great influence on the success of the innovation project (Qingyu & William 2001). As ideas are generated in the front end, this is both the most troublesome and the weakest part of the innovation process and, at the same time, the phase that represents the greatest potential (Reid & de Brentani 2004). The outcome of the FFE is a well defined concept,

Figure 1: What is The Fuzzy Front End?
Howe School of Technology Management

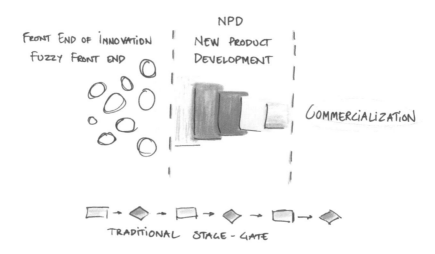

clear development requirements, and a business plan aligned with the corporate strategy (Kim & Wilemon 2002).

According to Moneart et al. (1995), a firm formulates a product concept and determines whether or not it should invest resources to develop the idea through the FFE. Based on the process developed by Cooper (1988), Murphy and Kumar (1997) define the pre-development stages as consisting of idea generation, product definition, and project evaluation.

In this chapter the FFE is defined as (Kim & Wilemon 2002):

> "...the period between (when) an opportunity is first considered and when an idea is judged ready for development" (p. 269).

The FFE phase, thereby, includes the development of the concept but not the concrete product.

As shown by (Murphy & Kumar 1997), the management of the FFE in intra-firm settings is essential, and unsuccessful management of this phase can have considerable consequences. If the project enters the development phase without

sufficient preparation, there is a high risk of project delays and budget escala-
tions (Kim & Wilemon 2002). Furthermore (Clark & Fujimoto 1991) point out
that engineering changes occurring late in the development are costly and time
consuming.

Table 2: Characteristics of the FFE phase in an intra-firm perspective
Adapted from (Kim & Wilemon 2002)

Factors	General Characteristics of the FFE Phase — Intra-firm Perspective
State of an Idea	Probable, fuzzy, easy to change
Features of Information for Decision-making	Qualitative, informal and approximate
Outcome (/Action)	A blueprint (/diminishing ambiguity to decide on whether to make it happen)
Width and Depth of the Focus	Broad but thin
Ease of Rejecting an Idea	Easy
Degree of Formalisation	Low
Personnel Involvement	Individual or small project team
Budget	Small/none
Management Methods	Unstructured, experimental, creativity needed
(Visible) Damage if Abandoned	Usually small
Commitment of the CEO	None or small

In the following sections, the factors Management Method, Formalisation and
CEO Commitment from table 2 will be considered in an inter-firm collaboration perspective. These have been chosen for further elaboration as the analysis
revealed considerable differences in this respect between the intra-firm and
inter-firm setting.

Case Description

The case started with KMD having introductory meetings with potential network partners at their respective company localities. The purpose of these meetings was clear to both parties involved. KMD would present the idea of a digital platform for B2C services within the energy area and wanted to find out whether the partner was interested in participating in the development of this project. At the end of each meeting, KMD invited interested partners to participate in an forthcoming meeting where all interested parties would be invited. Two months after the last introductory meeting was held, all interested network partners participated in a meeting held at KMD's meeting facilities. The purpose of this meeting was communicated as an opportubity to meet the other participants and to provide input to the concept development of the platform.

At the first meeting, which the director of KMD chaired, KMD presented their perspective on the digital platform. Their presentation included functions which should be incorporated into the platform as well as a flow chart illustrating how information flows in the system could be structured.

During the six hour long meeting at KMD, there was a lot of discussing and brainstorming concerning a wide variety of issues. The main topics were price, market potential and functionality of the product. The discussions were unstructured and the topics discussed came about as the participants put them forward. At the end of the meeting, the general assumption was that nothing new had arisen in regard to the functionalities of the platform and that most questions were still unanswered in regard to the market and the price.

In response, KMD suggested a second meeting to be held one month later where new participants with different backgrounds should be invited in addition to the present participants. The purpose of this second meeting would be creative thinking and idea generation concerning the platform.

The first meeting ended with KMD handing out questionnaires to the participants with the purpose of taking status as to which participants would still like to be involved in the development of the platform. The participating network partners expressed their wish to receive the minutes from the meeting and KMD agreed that minutes would be put on the website.

The second meeting was postponed four months and the minutes were not put on the website. Even though many of the participants had answered positively in the questionnaire they did not show up at the second meeting. A considerable group of participants did not think that there was a concrete outcome of the meeting and, therefore, they chose not to participate in the following meeting.

Case Analysis: Management Methods and Formalisation

The term *Management Methods* is closely linked to the term *Formalisation*. Accordingly both will be considered in this part of the analysis.

In the traditional intra-firm way of going through the FFE process, the loose idea or opportunity spotted is still an internal process within the company. In such an intra-firm perspective, a relatively unstructured management method is traditionally used. The work done in relation to the idea is often characterised by a very low degree of formalisation. Meetings are held without agendas and are done on an accidental basis or due to a coincidence of events. The management method is unstructured and encourages experimentation. This is a management method which supports the creative process and a desire to explore ideas and opportunities. Ideas that arise later become formalised projects in the NPD process or disappear without notice. Some survive in other projects and some are gone forever. In an inter-firm perspective, however, the lack of structure is an immanent issue to be addressed by the company.

As a focal firm invites network partners to participate in this stage of the innovation process they should consider that their partners would expect some kind of outcome. This is related to the issue of resources. The invited partners might need to travel in order to participate and they most certainly will need to spend time participating. Thus, it is not without expense to participate in the meetings. One thing is to meet informally at the site of your own firm, but it is quite a different thing to spend considerable resources preparing, travelling and participating in meetings with the sole purpose of providing potential innovation partners with valuable input.

According to our case and an analytical approach, it seems reasonable to claim that there ought to be a structure that benefits all partners stages of the FFE. An informal meeting in an intra-firm perspective is quite different from an 'informal' meeting in an inter-firm perspective. If participants do not perceive a tangible outcome they are likely to quit the network. On the other hand, too much structure in the concept that is to be developed will suppress creative inputs from the innovation partners.

Two levels of structure should be considered:

1. Structure of content: open vs. closed concept
2. Structure of work process: structured vs. unstructured

The structure of content is related to the thinking processes concerning the concept that is to be developed during the FFE. If the concept is presented as a closed concept with specified technologies, functionalities and information flows, the way participants will discuss the project will be within the structure of that concept. If the concept is presented as an open-concept with multiple alternatives regarding technology, functionality and information flows, the participants will be more likely to come up with creative input (Basadur et al. 2000). In the case in point, KMD presented the digital platform as a somewhat closed concept leading to a lack of innovative inputs from the participating partners concerning the concept.

The structure of process refers to the work process in relation to the development. Is there a clear purpose and agenda of meetings? Is is clear who the chairman is? Is the outcome of each meeting made explicit? The case has shown that, in the case of the first meeting in the network, the purpose of the meeting was clear but the outcome of the meeting was very unclear. Even though the participants explicitly asked for a tangible outcome in the form of minutes, such were never provided for them. Instead, the outcome was presented to them as an upcoming meeting which, in the end, was postponed four months.

Figure 2 illustrates the difference between going through the FFE phase in an inter-firm setting vs. an intra-firm setting towards the goal of the FFE phase – a well defined concept, clear development requirements, and a business plan aligned with the corporate strategy.

In the intra-firm setting, the process can be unstructured from the beginning and gradually become more structured as the personnel involvement becomes clearer and the concept takes form (Jongbae & David 2002; Kim & Wilemon 2002). It can also start as a somewhat closed concept developed by a small group or an individual and then be challenged by colleagues and as a result become open.

Figure 2: Two Levels of Structure in FFE Concept Development
Based on Table 2 and Case Analysis

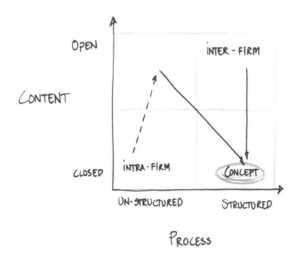

Managerial Implications

In the inter-firm setting, it appears central that the process takes off with an open concept where the collaboration is based on a structured process. The interaction between participants is limited and the structure should support creative inputs and new perspectives on the concept.

The challenge in regard to structure, of going through the FFE in an inter-firm setting is, thus, to provide successive tangible output in an intangible process. Balancing the two levels of structure ensuring that the concept is presented as open, so that creative input from participants is elucidated and having a clear structured work process that explicitly reveals output and progress in the process is paramount. As there can be no use of direct fiat in such inter-firm settings, the focal firm needs to structure the development work process and the creative thinking process in a way that provides tangible output and allows the organisation of knowledge and, thereby, new ideas (Brown & Duguid 2000). The risk is that partners will give up their commitment if they do not see a continuous development and progress in the process.

Case Analysis: CEO Commitment

The top managers of a firm must consider the role that he or she plays in the FFE phase. The attention and decisions on the level of commitment are clearly linked to the management methods, formalisation and structure considerations discussed above. If the degree of formalisation of the FFE phase is high, the CEO is more likely to be actively involved than if it is a bottom-up, team-based, unstructured phase.

As stated in table 1, the level of top management involvement in the FFE phase in an intra-firm perspective is generally limited or even not existing. The unstructured experiments by the individual employees do not imply involvement by senior managers. Mid-level managers might be involved in terms of letting the employee(s) have some hours per week to carry out their experiments but, otherwise, the intra-firm FFE phase does not require a high degree of management involvement.

The issue of CEO involvement in the inter-firm FFE phase is quite different. As a point of departure, the top managers are part of the idea selling process: They have to convince the potentially participating network partners to take part in the project. The fact that the CEO is actively involved can contribute to stress the importance that the focal firm is committed to the outcome of the specific project and the inter-firm collaboration as such. One should keep in mind, that the collaboration is still not formalised and ,thus, can be characterised as a network where the CEO has no direct power.

The case analysis has revealed, however, that the active involvement and commitment of the top managers is a double-edged sword. If the CEO in an inter-firm FFE phase is too committed, it might make the other participating organisations nervous about a potential bias in the distribution of the benefits generated from the collaborative effort.

The director of KMD addressed this balance in numerous discussions with his colleagues and the researchers that have been involved. At the first meeting, he chose the rather active role because he expected this to enhance the involvement of the participating organisations. At the second meeting, he outsourced the chairing role to an external process consultancy in order not to be too dominant, and thus hampering the innovative processes.

Managerial Implications

The case analysis illustrates the delicate balance between a high level of commitment from the top manager to convince the potential partners to get involved, on one side, and the fear of scaring the partners away from the project, on the other. Thus, the involvement of the top manager is an exercise in understanding the preferences of the participating organisations in terms of showing commitment, on one side, and not being too eager, on the other.

Conclusion

This chapter has discussed the differences of going through the FFE phase of the innovation process in an intra-firm versus an inter-firm setting. In respect to collaboration form, the latter is characterised as a *network*, where no direct power can be employed, as opposed to an intra-firm setting which is characterised as *hierarchy*. Through an in-depth case study analysis, the focus has been on two main differences – namely the *management methods* (formalisation) and the *CEO-commitment*. In regard to *management methods*, the analysis showed that, in an inter-firm setting, two levels of structure should be considered:

1. The structure of content: whether the concept is formulated as an open or a closed concept.

2. The structure of the work process: whether there is clear purpose of the meetings and the outcome of each meeting is made explicit.

CEO and top management commitment in an inter-firm setting proved to be a balance between using this commitment to stress the focal companies' commitment to the project, in order to get the network partners to participate and not becoming too committed, as this could result in participants leaving the project.

Further Research

Further research could be aimed at a better understanding of why organisations join networks of collaborative innovation. The present research has shown that having insight in the expectations and motivations of participating organisations in a network collaboration setting is of great value to the focal company. Developing a methodological approach for acquiring such knowledge should be of high priority to both academics doing research within the field of network dynamics and practitioners balancing the structure of content, work process and CEO-commitment on a daily basis.

Acknowledgements

This chapter is based on work carried out under the NEWGIBM-project funded by The Danish Ministry of Science, Technology and Innovation. The authors would like to thank the Ministry and the case study company for the generous degree of participation in exploring innovation carried out in network collaboration.

References

- Bart Nooteboom 2003, *Inter-firm collaboration, networks and strategy; An integrated approach* The Hague.

- Basadur, M., Pringle, P., Speranzini, G., & Bacot, M. 2000, "Collaborative Problem Solving Through Creativity in Problem Definition: Expanding the Pie", *Creativity and Innovation Management*, vol. 9, no. 1, pp. 54–76.

- Brown, J. S. & Duguid, P. 2000, *The Social Life of Information* Havard Business School Press.

- Clark, B. K. & Fujimoto, T. 1991, *Product development performance: strategy, organisation, and management in the world auto industry / Kim B. Clark, Takahiro Fujimoto* Harvard Business School Press, Boston, Massachusetts.

- Cooper, R. G. 2005, *Product Leadership: Pathways to Profitable Innovation* Basic Books.

- Cooper, R. G. & Kleinschmidt, E. J. 1987, "Success factors in product innovation", *Industrial Marketing Management*, vol. 16, no. 3, pp. 215–223.

- Eisenhardt, K. M. 1989, "Building Theories from Case Study Research", *Academy of Management Review*, vol. 14, no. 4, p. 532.

- Eisenhardt, K. M. & Graebner, M. E. 2007, "Theory Building from Cases: Opportunities and Challenges", *Academy of Management Journal*, vol. Vol. 50, no. No.1, pp. 25–32.

- Faems, D., Van Looy, B., & Debackere, K. 2005, "Interorganisational Collaboration and Innovation: Toward a Portfolio Approach*", *Journal of Product Innovation Management*, vol. 22, no. 3, pp. 238–250.

- Flyvbjerg, B. 2006, "Five Misunderstandings About Case-Study Research", *Qualitative Inquiry*, vol. 12, no. 2, pp. 219–245.

- Hagedoorn, J. 2002, "Inter-firm R&D partnerships: an overview of major trends and patterns since 1960", *Research Policy*, vol. 31, no. 4, pp. 477–492.

- Jongbae, K. & David, W. 2002, "Strategic issues in managing innovation's fuzzy front-end", *European Journal of Innovation Management*, vol. 5, no. 1, p. 27.

- Kim, J. & Wilemon, D. 2002, "Focusing the fuzzy front end in new product development", *R&D Management*, vol. 32, no. 4, pp. 269–279.

- Murphy, S. A. & Kumar, V. 1997, "The front end of new product development: a Canadian survey", *R&D Management*, vol. 27, no. 1, pp. 5–15.

- Powell, W. W. 1990, "Neither market nor hierarchy: Network forms of organisation", *Research in Organisational Behavior*, vol. 12, pp. 295–336.

- Powell, W. W., Koput, K. W., & Smith-Doerr, L. 1996, "Interorganisational collaboration and the locus of innovation: Networks of learning in biotechnology", *Administrative Science Quarterly*, vol. 41, no. 1, p. 116.

- Qingyu, Z. & William, J. D. 2001, "The fuzzy front end and success of new product development: A causal model", *European Journal of Innovation Management*, vol. 4, no. 2, p. 95.

- Reid, S. E. & de Brentani, U. 2004, "The Fuzzy Front End of New Product Development for Discontinuous Innovations: A Theoretical Model", *Journal of Product Innovation Management*, vol. 21, no. 3, pp. 170–184.

- Tidd, J., Bessant, J., & Pavitt, K. Managing Innovation: Integrating Technological, Market and Organisational Change, 3rd Edition, 17.2005. Ref Type: Generic

- Yin, R. K. 1994, *Case study research: Design and methods* Sage Publications: London.

Chapter III.4

NEW Global Business Models – What Did The NEWGIBM Cases Show?

Peter Lindgren

&

Yariv Taran

Abstract

Global innovation is making it increasingly difficult for companies to innovate individually. Therefore, in addition to developing their core innovation competence base individually, enterprises are drawn towards more open and network-based innovation, particularly when the core knowledge and competences needed for innovation are not available in-house. The customers, suppliers, consultants and even competitors are presently recognised as the main potential sources of external innovation competences. This is not a new phenomenon but it is important as these functions are integrated actively in tne innovation process.

Collaboration helps companies find and open new business models, and obtain a better understanding of what is expected from them in the business model. The NEWGIBM cases are clearly an example of the above mentioned practices.

This chapter examines the business models that the NEWGIBM case companies were innovating and focuses on the challenges related to such innovation. The aim of this chapter is to reflect theoretically on the business models of the NEWGIBM case companies. It answers the questions: a) What did these models show? and b) How can they be related to the NEWGIBM framework of business models and are they in any concern new and global business models?

Keywords: Business models of the NEWGIBM case companies, new business models, cross functional innovation.

Introduction

Many of the NEWGIBM case companies are finding themselves increasingly tied to other firms - sometimes referred to as the extended enterprise or the interactive firm (Johansen and Riis 2005). The challenge each of them faces is:

to adjust their own business model to their network partners 'to meet their partners' core competences and innovate on the basis of them to create a new and joint business platform for collaboration and innovation to become global and create a global business model

The case studies show that it is useful to study the business model concept not only at the level of the individual firm, but also at the horizontal value chain level (supplier, company, distributor, and customer). It is also necessary to analyse the whole network surrounding the firm, both vertically and a horizontally as well as introduce more levels for analysis. This could be e.g. the whole chain of customers or suppliers (Chan 2005).

In all the NEWGIBM case-studies, it was clearly noticed that, in order to successfully innovate, it was necessary for the companies to regulate their individual BMs to the overall network partner's skills and competences. The level and degree of collaboration, the individual BM modification, the success rate of innovation and the network-based BM formations were all considered to be relatively diverse in each case study.

When companies regulated their individual BMs to the overall network partner's BM a new integrated BM was created. During the research project it became evident that the following questions were important for the understanding of innovating BMs. What are the characteristics of these BMs and which challenges and opportunities do they offer? How do these new BMs influence the skills and competences of the individual companies? Table 3 illustrates the answers to these questions.

What Did the Cases Show?

The NEWGIBM project comprehended eight networks innovating eight different business models. The following sections will comment on each of the eight networks.

The Intelligent Utility Case (IUC)

In the Intelligent Utility Case, one core company steered the network towards a network-based business model. The challenge was to gain the attention of the other network partners and to capture their interest in shaping a new model. The network partners were spread all over Denmark, and both the business model that emerged and the joint product innovation activity were new to the industry. Hence, the main effort here was to exploit the competences of each network partner for the purpose of leveraging the entire network's capability to compete successfully nationally and, later on, also globally. Furthermore, it was discovered that providing both the triggers and the drivers for getting the network partners to involve themselves in the innovation process was a main criteria for success. The success criteria were both short and long-term as well as financially and/ or non-financially oriented. It was also found that the success criteria could be related to the perceived value and cost of products and relationships.

The intelligent utility case showed a BM that was reworked into a new BM. The customer function was new and the BM was constructed as a network based, cross-functional and open BM. It was not possible to talk about any type of diversification to any of the network partners. The focus on success criteria was both short term and long term based. The BM was not focused on being global. The BM was highly supported and enabled by ICT.

The Master Cat Case (MCC)

In the Master Cat Case, network relations included local as well as regional (Denmark-Norway) partners, and the new ferry is considered to be a new network and product/service to the region. The interesting conclusion that can be drawn from this case study concerns the influence that the nature of the network has on the business model value equations of the individual partners. In the more 'contract-free' Hanstholm Network, the value sets and the business models of the participants were more flexible and diverse. The more the network partners were locked into the network-based business model, like in the Norwegian-Danish joint venture, the more committed they were to ensuring the overall success of the specific task chosen. Consequently, individual partners adjusted their business model to better fit the network level model.

The Master Cat Case showed a BM that was not new to the world – the low cost transportation BM was copied from the flight industry (Ryanair, Easyjet e.g.). Customer function, customer group and customer technology were known. The perceived values and cost, however, were surprisingly different from what the network partners had first believed, because private customers from Norway were willing to travel as far as 6 hours to the new ferry to save 90 minutes of travel time. Transportation companies also liked using the ferry because they could use a better road and saved time to the German border. By doing so they could reduce their transportation costs by 25% as compared to using competing ferries.

The customer function, customer groups and technology were well known. It is not possible to talk about diversification in any of the networks. The BM was not open. It was developed in a rather closed network due to the BM conditions. The BM could not be characterised as truly new or as global although it is believed that the concept could be copied in other travel companies as well as in other lines of business. The business model was not particularly based on ICT, however the ICT enabled a cost effective sales channel – namely via internet booking and sales of tickets.

The SKOV Case (SC)

Case has not been allowed to be published.

The Tricon Case (TC)

Case has not been allowed to be published.

The Danfoss Industry Service Technology Case (DIS)

In the Danfoss case the unit of analysis was the 'Industry Service Technology Centre', which provides internal services for Danfoss's three major divisions, and also external services for other non-Danfoss customers (companies). Although the centre has the potential to provide a wide range of services to suit diversified (global) companies' service requirements, the centre is presently operating only in the vicinity of the company's headquarter location - the region of southern Denmark. Here an interesting question arises: Why does such a centre, that is part of a well-known and worldwide company branded with high quality products and services, have a situation with zero internationalization? After all the 'Industry Service Technology Centre' is able to deliver high quality service, with incredibly competitive prices and, as a consequence, create satisfied and loyal global customers.

One of the most important findings in this case, is to learn to better appreciate the importance of the ICT usage when one tries to become global. It seems that human capital, organisational culture, and the ICT skills of employees play a vital role in the internationalisation process of firms. According to the case study results, Danfoss is dominated by a classic engineering culture - a culture that focuses on mechanical or very tangible aspects of doing business. An unfortunate consequence of this organisational cultural trait is a lack of virtual aspects offered to the market. This applies to both the services they offer and to facilitation possibilities that virtually are there representing among other things i.e. information technology, data bases, fast communication, customer cooperation, test facilities etc. 'Virtually' means that the facilities exist when the customers need the facilities and when there is a customer task to be solved with the facilities – otherwise the facilities are not existing. The employees of the company also face internationalisation barriers such as language and cultural differences.

Another challenge in the 'Industry Service Technology Centre', has been the relatively long waiting time for the testing of a customer product. Reducing this customer waiting time is a key success factor to consider. For that reason, a NEWGIBM network has been proposed, namely, to develop an Industry Service Technology portal solution for the testing of products. This includes a booking system for standardised 'service packages' in order to reduce waiting time, as well as to provide better Industry Service Technology relationships.

In the Danfoss case, the customer function and customer technology were mostly unknown while the customer group was known. There was no diversification and the BM was not open. The BM was developed in a closed network but the company desires to develop to the business model in an open network in the future. The BM could be characterised as partially new and, although focused on the global market, not a global BM in a short-term perspective. The business model was particularly based on ICT. The value and cost and perceived value and cost by the customers must be specified more in details in order to determine if there was anything new about the BM. The BM was focused on the global market.

The Nano Solar Case (NSC)

The Nano Solar Case, 'ISO Paint Nordic A/S was a company specialising in the manufacturing and developing of facade paints and impregnations. They had decided that, in order to meet their declared corporate strategy of continuing to be a top quality, environmentally friendly, and an industrial pioneering enterprise, they should therefore (over-proportionally) promote their scientific, application-oriented, research and development, and cooperate with renowned suppliers, dealers, professional roof coaters and other leading institutions.

Due to the new 'R&D in network' strategy, 'Isopaint' has gained qualitative knowledge as well as additional core competence for exploring new innovation possibilities, particularly with emphasis on a Nano solar paint that was in focus. The innovation project and the new business model was focused on the global market and intended to be developed in a large and diversified national network. When launched, the benefits were primarily the ability to implement the project itself. Additionally, energy savings, energy stability, customer satisfaction and security benefited. Furthermore, the payback time on the investment to the customer could be diminished tremendously. Finally, the company could meet its own declared corporate strategy and continue being an industrial pioneering enterprise.

This case study is a good illustration of how a company identifies an opportunity, while also recognising its own inability in trying to exploit it without outside assistance. Therefore, they chose to externalise the idea and business model - to share it with chosen network partners that could assist the company in utilising the idea more successfully - transforming theory into practice via using the network partners' combined core competences and skills.

The customer function, customer technology and customer group were mostly unknown because the product had never been seen before on the market. It is, indeed, possible to talk about diversification in this BM, because the BM and the related products and services will be applied to other markets – the automobile market, farm building industry and building industry among others. Further, the BM is not related to Isopaint's existing BM – it includes new customer functions and new customer technology. The customers, both end users and distributors could, however, be the same. The BM is rather open. The product will be developed in a somewhat closed network initially but network partners wish to develop to a more open network. The BM could be characterised as new and focused on the global market – a global BM focused both on a short-term and long-term perspective. The business model is mainly based on ICT along with advanced technology from the parent companies (nanotechnology, power energy e.g.). The value and cost and perceived value and cost by the customers have to be specified in more detail. Certain obvious values and cost can be registered immediately as well as some perceived value and cost can also be seen. The BM is focused on the global market.

The dixiTel Case (dX)
Case has not been allowed to be published.

The Smart House Case (SH)
Finally, in the 'Smart House'case, the new (local) network was inhomogeneous (both vertically and horisontally). It was actually very diverse in terms of the sectors (cross-industrial/diagonal borders), the partners represented and the products and services they produced. This was beneficial to the radical innovation or "Blue Ocean" strategy (Chan and Mauborgne 2005) pursued. The greatest challenge was (and still is) to get the various partners interested in forming such a network. This interest is strongly related to the network partners' value equation or 'what is in it for me?' For that reason, the entire Smart House concept was directed by the North Jutland county in order to avoid suspicion and mistrust (that only one company's interest was served), and to support the network in the innovation process.

The value equation in the interaction between the network partners was extremely important to understand in order to drive the innovation project forward. The network partners were playing with and/or seeing very different business models – some very short-term oriented, others much longer-term based (research, learning, know-how or business in the future); some finance focused and others predominantly non-financial. The understanding of the value equation of the whole chain of customers was important in order to understand the potential of the business model.

The customer function, customer technology and customer group were mostly known beforehand. There are, however, certainly elements that were unknown because some of the products had never been seen on the market before. There was no diversification because the BM and the related products and services were brought into known markets. The BM is somewhat open. The product will be developed in a rather closed network initially but there is a desire for the network to develop into a more open network. The BM could be characterised as new and focused on the global market – a global BM focused both on a short-term and a long-term perspective. The business model is primarily based on ICT along with advanced technology from other lines of business. The value and cost and perceived value and cost by the customers have to be specified in more details. Certain obvious values and cost can be noticed immediately as well as some perceived value and cost. The BM is focused on the global market.

Reflection and Discussion

The cases clearly point to an observation that was already mentioned earlier in chapter I.2 'The Theoretical Background of Business Models Innovation', namely that the business model concept cannot be considered only as one holistic BM unit design related to one focal company. The BM has to be seen as a holistic and interrelated business platform, where each network partner is adjusting his own BM formation to meet other partners' BM formations within the network. This will eventually lead to a Network-Based BM formation that ends up in comprising all network partners' core competencies, skills and BMs. Therefore, the new business model has a different organisational structure and value chains than literature has described until now, and this is by definition a Network-Based BM.

In the table below it is shown how the Danish case companies worked with the new business model concept. Naturally, the complexity of explicitly detailing each building block is very time consuming. In order to simplify things the fields value and process propositions have been incorporated in one field. This is also the case for the cost structure and revenue model building blocks in each case study.

Table 1: Building Blocks of the New (network based) Business Model

New Global Business Model Building Block	Intelligent Utility	Master Cat	Smart House Smart City
Value and Perceived Value Proposition	Network partners are sharing common values and processes in order to develop a new intelligent energy steering concept for private houses.	**Norwegian Group**: Establish a profitable ferry business.	Network partners are independent and equal partners sharing common interest – to develop a new smart house business platform focused on the global market
Process Value and Process Value Proposition		**Danish Group**: establish a profitable ferry and Tourism business to develop the entire Hanstholm region.	
Target Customer	All private houses	Both groups share the same target customer: ferry passengers and cargo customers. The Hanstholm group sees this as a possibility to attract other customer groups to the area like skiing tourists, other cargo companies e.g.	New buildings of private houses, institutions e.g. and customer with interest of smart building components Established living houses with interest of smart building, smart building components e.g.
Distribution Channel (Physical, Digital, Virtual)	Virtual	Physical	Physical, digital and virtual

New Global Business Model Building Block	Intelligent Utility	Master Cat	Smart House Smart City
Relationship	Steered by an intelligent device and software (virtual relationship)	**Norwegian Group**: Customers: Passengers on the ferry Cargo customers **Danish Group**: Customers: 'On' and 'Off' the ferry	Steered by both physical, digital and virtual relationships
Value and Perceived Value Configuration	Diversification of values and perceptions (short vs. long term perceptions)	**Norwegian Group:** Shared common values **Danish Group**: Diversification of values and perceptions	Diversification of values and perceptions (Short vs. long term perceptions)
Core Competence	Different partners' core competences: Electricity, telephone, gas, water, and software programs.	**Norwegian Group**: Strong network ties by contracts (venture capital), harbour resources. **Danish Group**: Strong network ties by trust, harbour resources.	Different partners' core competences: Building, architecture, building research, IT, environment, security, building material.
Partner Network	National network	Local/ regional network	Local/ cross regional network
Cost Structure Revenue Model	Some network partners are finance focused, others predominantly non-finance focused. Short and long term financial focus.	Short and long financial focus.	Some network partners: Norwegian group, Hanstholm Havn finance focused Other partners: Predominantly non-finance focused

Table 2: Building Blocks of the New (network based) Business Model

New Global Business Model Building Block	Tricon Data
Value and Perceived Value Proposition	Involved in development activities together with a number of clients. Developed a portal solution and a 'configurator' as a platform for communication and collaboration concerning the project.
Process Value and Process Value Proposition	
Target Customer	Electrical, electronical and data systems companies
Distribution Channel (Physical, Digital, Virtual)	Physical, digital and Virtual channels
Relationship	Until now mainly physical and digital - lately also (global) virtual channels.
Value and Perceived Value Configuration	Diversification of values and perceptions. (Short vs. long term perceptions). The lack- of experience, trust and timing has caused a setback to the network potential in leveraging the innovation success.
Core Competence	Relied on their own competencies – perhaps including the help from subcontractors.
Partner Network	Global physical, digital and virtual partner network.
Cost Structure	Mainly financial and long term focus.
Revenue Model	

Table 3: Building Blocks of the New (network based) Business Model

Value and Perceived Value Configuration	Share common values with the 'mother' company (Danfoss).	All network partners see this as a possibility to innovate new values and perceived values to their customers and their product, business model.
New Global Business Model Building Block	Danfoss Industry Service Technology	Nano Solar Case
Value and Perceived Value Proposition	Makes 'Industry Service Technology' able to deliver high quality to a very competitive price thus creating satisfied and loyal customers. The wIndustry Service Technology has different focused offerings. They perform technical testing, analysis of various themes, consulting services, calibration, failure analysis, small series production, prototypes and information services, etc.	'R&D in network' - decided that in order to meet their declared corporate strategy to continue being a top quality, environmental friendly and industrially pioneering enterprise, they should promote their scientific, application-oriented research and development, and cooperate with renowned suppliers, dealers, professional roof coaters and other leading institutions.
Process Value and Process Value Proposition		

Value and Perceived Value Configuration	Share common values with the 'mother' company (Danfoss).	All network partners see this as a possibility to innovate new values and perceived values to their customers and their product, business model.
Target Customer	Aiming globally, unfortunately now they operate only locally - close to the testing facilities.	Not just their own present target customers, but also all network partners' (different) target customers: ISOPAINT , POWER LYNCH, DANFOSS Aalborg University Nano research community Energy research community Solar energy research community New global business model research community University in Magdeburg Nano research community Other industry in need of research Other EU – research communities Farbe Industrie Dresden: Roof Painting companies Other industries – energy industry, parts of building industry other than those served to day Alpin BAU, SKOV A/S All network partners see this as a possibility to attract 'other customer' groups to the area of interest.
Distribution Channel (Physical, Digital, Virtual)	Mainly physical.	Physical, digital and virtual channels

Value and Perceived Value Configuration	Share common values with the 'mother' company (Danfoss).	All network partners see this as a possibility to innovate new values and perceived values to their customers and their product, business model.
Relationship	Close and physical channels.	All network partners have different physical, digital and virtual relationship All network partners see this as a possibility to enter both relationship with partners within the network and new relationships with partners connected to the project during the innovation period
Core Competence	Relied mainly on their own 'In house' (Danfoss) competence	ISOPAINT: Specialised in the sector of manufacturing and developing facade paints and impregnations. POWER LYNCH, Danfoss: Specialised in the sector of power energy equipment. Aalborg University Nano research community Energy research community Solar energy research community NEWGIBM research community University In Magdeburg Nano research commun ity Paint research community Farbe Industrie Dresden: Specialised in the sector of paint and chemical production and development. SKOV: Market leaders in the field of climate control and production monitoring for animal agricultural production.

Value and Perceived Value Configuration	Share common values with the 'mother' company (Danfoss).	All network partners see this as a possibility to innovate new values and perceived values to their customers and their product, business model.
Partner Network	Relied mainly on their own 'In house' (Danfoss) competences.	Global physical, digital and virtual network partners
Cost Structure	Short and long term financial focus.	All network partners have different cost structure and revenue models. Some network partners are finance focused, others predominantly non-financially oriented. Short and long term financial focus.
Revenue Model		

Our findings have shown that very few case companies are focusing on innovating global business model and the discussion of the global potential of their network at the initial part of the innovation project is very little. They normally choose to innovate and act locally (Intelligent Utility, Master Cat, Tricon) and then afterwards go for the global market. This affects the level and potential of their innovation both in a long term and short term perspective. We believe that issues about globalization have to be 'built-in' from the very start of the innovation process and have to come more into focus in the innovation process. Otherwise the innovation for the global market will be more difficult and maybe not even commence and Danish companies will loose their innovation potential. The innovation process has often to be started again or revised because of the lack of globalization in the innovation process. Often there is something missing in the innovation process to have the BM rolled out and fit for the global market.

The NEWGIBM BM showed a variation of newness in BM related to our framework. In the table below we tried to list up and analyse the specific characteristics of the BM related to the newness of the BM.

As can be seen very few of the BM can be classified as completely new BM. They have elements on every characteristic that could touch the definition of new. In

Table 4: Specific characteristics of the BM

Characteristics	'Old' Business Model Formation	A New Business Model	IUC	MCC	TC	DIST	NS	SH	I	I/R	R
1. Innovation	Incremental Product, Process. Position and Paradigm Innovation (I)	Radical Product, Process. Position and Paradigm Innovation (R)	(R)	(I)	(I)	(I)	(R)	(I/R)	3 (50%)	1 (17%)	2 (33%)
2. Markets											
Customer Functions (CF)	Familiar Customer Functions (F)	Unfamiliar Customer Functions (U)	(U)	(F)	(F)	(F/U)	(F/U))	(F/U)	2 (33%)	3 (50%)	1 (17%)
Customer Group (CG)	Familiar Customer Group (F)	Unfamiliar Customer Group (U)	(F)	(F)	(F)	(F)	(F/U)	(F/U)	4 (67%)	2 (33%)	
Customer Technology (CT)	Familiar Customer Technology (F)	Unfamiliar Customer Technology (U)	(F)	(F)	(F/U)	(F/U)	(U)	(F/U)	2 (33%)	3 (50%)	1 (17%)
3. Diversification	Product and Market Development Integrative Growth (-D)	Diversification (D)	(-D)	(-D)	(-D)	(-D)	(D)	(-D)	5 (83%)		1 (17%)
4. Network	No Networks nor Familiar Networks (Physical, Digital, Virtual Networks) (F)	Unfamiliar Networks (Physical, Digital, Virtual) Networks (U)	(F/U)	(F/U)	(F/U)	(F)	(U)	(F)	2 (33%)	3 (50%)	1 (17%)

Characteristics	'Old' Business Model Formation	A New Business Model	IUC	MCC	TC	DIST	NS	SH	I	I/R	R
5. Competences	Based Solely on Internal Competence and Familiar Competences (E)	Based on Cross Functional Competences - Across Networks (U)	(U)	(F/U)	(F)	(F)	(U)	(U)	2 (33%)	1 (17%)	3 (50%)
6. BM	Individual, Closed (IC)	Network-based, Open (NO)	(NO)	(IC)	(IC)	(IC)	(NO)	(NO)	3 (50%)		3 (50%)
6. Success Criteria	Short Term (ST)	Long Term (LT)	(ST/LT)	(ST/LT)	(ST/LT)	(ST/LT)	(ST/LT)	(ST/LT)		6 (100%)	
7. Innovation Technique	Innovation Based Mainly on Incremental Product, Process Innovation (PI)	Innovation Based on Business Platform Innovation With the Possibility to Innovate a Variety of Products/Services Processes e.g. on the Platform. Open Innovation (BMO)		(PI/BMO)	(PI/BMO)	(PI/BMO)	(BMO)	(BMO)		3 (50%)	3 (50%)
8. Global	Innovation and Business Model Based on National Markets (-G)	Innovation and Business Model Based on Global Markets (G)	(-G)	(-G)	(G)	(G)	(G)	(G)	2 (33%)		4 (67%)

the radar diagram beneath we have put in the 6 case companies related to the New BM definition.

The radar diagram below shows the 6 case companies related to all the dimensions of the new BM definition
As can be seen it is only one BM in the diversification area (Nano Solar Case) that enter the radical new BM dimension.

Very few NEWGIBM case BMs could be characterised as Open BMs - although some of the case companies wanted them to be open in the future. It seems as if it is quiet difficult to keep the innovations process open to many network partners and the case companies are not familiar with this innovation terminology and method.

The authors' proposal for adding new elements to the existing building blocks, the perceived value set and the perceived cost set, was verified in the Nanosolar, the Mastercat, the Smarthouse and the Skov cases. However it was documented that perceived value and cost are very different to each network partner. During the present research project it was found that one of the main challenges to NEWGIBM innovation is to understand and to find each network partner's business model value/cost and perceived value/cost set. Between these value sets

Figure 1: Radar Diagram of NEWGIBM BMs

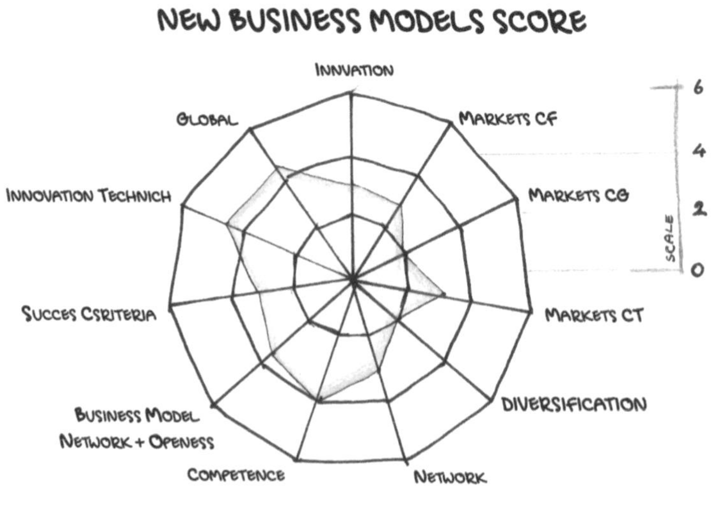

NEW BUSINESS MODELS SCORE

and perceived value sets the drivers for entering, starting and stopping the innovation process for each partner could be identified.

It was also proved that the new (local / global) network ties were not necessarily characterised by industrial homogeneity (vertical or horizontal), but was characterised by a focus on large diversification (cross-industrial borders) on business models and their related products and services – for the purpose of pursuing radical innovation ('Blue Ocean') possibilities and developing real new business models.

The authors found that a new global business model cannot be considered only as one holistic unit design related to one focal company, but rather as constructed by many different (Meta) business models related to networks with many network partners. The new global business model has a different organisational structure and value chain, and it is by definition a network-based BM. Furthermore, the global orientation of the business model needs to be incorporated in the innovation process at the very beginning of the innovation process.

The BM has a strong emphasis on innovating in networks, both with commercial and non commercial methods and purposes. Furthermore, it was established that it is necessary to elaborate on the current success criteria (building blocks) of the 'plain' business model, and to push towards a New Global Business Model Design, where leaders could include also short and long term global success criteria, which are not necessarily financial driven alone, but can also be non financial driven in order to develop a sustainable business model orientation in the network.

As can be seen innovating new global business models is much more complex than 'a simple product (Cooper 1996) or process innovation (Davenport 1996)'. This will be commented further on in chapter III.7 'Successful Implementation of Global BM Innovation'.

Conclusion

During the present research and case studies, it was established that the authors' hypothesis to add new building blocks and to improve existing building blocks could be verified in the cases. What has been found up till now on the BM characteristics and specifically with emphasis on what Osterwalder described as the 'innovation process' of a BM has to be developed further as regards to business models in a network context.

Most of the BMs could not be characterised as Global BMs and the network partners in the innovation process did not even think of going global with the BM at the initial phase of the innovation process. It can not be verified through the cases why this is so but innovating for the global market adds ano°ther complex dimension to the innovation process. This could be an important reason.

The business model is no longer related to only one focal company but is related to the whole network and to all the network partners behind the new global business model with a chain and variety of business models from both up and down stream network partners. These findings are, as can be seen, very different to previous research on the concept of business models.

The new global business model is 'an integrated business model' where each network partner delivers a part to 'the collective business model'. Each network partner has his own individual business model that is integrated into a collective business model structure when implemented.

The present findings have also shown that New Global Business Models are in most cases based on or enabled by ICT. Some of the NEWGIBM cases had network partners whose success criteria were long term oriented and not related to the financial success criteria at all, for example: self-learning purposes, research without a financial output, political and environmental interests, et cetera.

Finally it was shown that very few of the BMs were open and to be qualified as really new business models. The network partners wanted the BM to be open and global but in a long term perspective.

References

• Chan, W. K. and R. Mauborgne (2005), Blue Ocean Strategy, How to Create Uncontested Market Space and Make Competition Irrelevant, Harvard Business School Press.

• Cooper RG; Kleinschmidt EJ, Journal of Product Innovation Management, November 1995, vol. 12, no. 5, pp. 374–391(18).

- Davenport T. H. (1993), Process Innovation: Reengineering Work through Information Technology, Harvard Business School Press,

- Johansen, J. & J.O. Riis (2005), 'The Interactive Firm - Towards a New Paradigm: A Framework for the Strategic Positioning of the Industrial Company of the Future', International Journal of Operational and Production Management, Vol. 25, no.2, pp. 202–216

Chapter III.5

The Strategy Concept in Light of the Increased Importance of Innovative Business Models

Anders Drejer
Kean Sørensen
&
Peter Lindgren

More than Just a Blueprint

Business model change and alternative business models compete for the same resources, competencies and customers. These are the cruel realities of management that cannot be neglected. This also means that the notion of strategy and strategic management cannot be neglected as an integral part of what managers do – and ought to do. In terms of BMs, strategy is about how the BM of the organisation is formulated, executed and changed. Strategy is the dynamic aspect, whereas the BM is the formulation of the status of the business.

In other words, strategy is about ways to make the organisation a winner through a competitive BM. Strategy is about survival in fierce competition and this chapter is about strategy. In this section, the authors will investigate the notion of strategy and strategic management seen in relation to the companies participating in the NEWGIBM project. They will also discuss some of the challenges – internal or external - that serve to push the limits of strategy. Finally, the key concepts of strategy will be defined in the light of today's increased importance of BMs.

The Potential of Business Models – Why Bother?

'There is something rotten in the state of Denmark'. In the past few decades, technological changes have become more frequent, market segmentation has increased considerably and global competition has made it more difficult to sustain a competitive position (Zeleny 2007). There are several reasons for this and looking back in history will give more of an understanding of the trends in the environment and how these trends have put pressure on companies. According to Drejer (2007), organisational forms have undergone profound changes through the past few decades. There is not merely one single cause for these changes. Rather, a number of underlying causes can be found, such as new market factors, more rapid technological changes and global competition. This has increased the importance of product development as well as underlined the importance of a renewal and an extension of existing product portfolio. This development is illustrated by the model in figure 1.

The business environment of many organisations has undergone radical changes since 1960s and successful companies have had to adjust to these changes in

Figure 1: The Evolution of Market Demands and (Required) Competencies of Firms
Boluijn and Kampe 1994

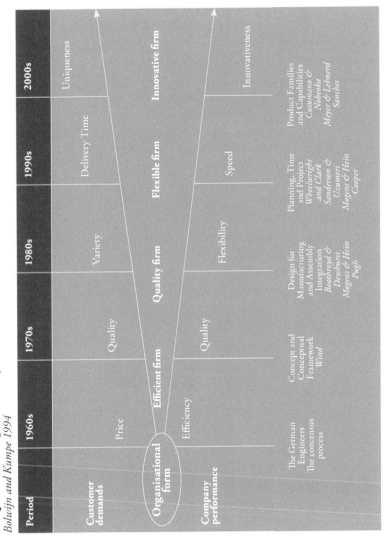

Period	1960s	1970s	1980s	1990s	2000s
Customer demands	Price	Quality	Variety	Delivery Time	Uniqueness
Organisational form	Efficient firm		Quality firm	Flexible firm	Innovative firm
Company performance	Efficiency	Quality	Flexibility	Speed	Innovativeness
	The German Engineers The concensus process	Concept and Conceptual Framework *Wind*	Design for Manufacturing and Assembly Integration *Boothroyd & Dewhurst Mogens & Hein Pugh*	Planning, Time and Project *Wheelwright and Clark Sanderson & Uzumeri Mogens & Hein Cooper*	Product Families and Capabilities *Cusumano & Nobeoka Meyer & Lehnerd Sanchez*

order to survive (Zeleny 2007). In the 1960s, successful companies focused on effectiveness by means of hierarchical organisational structures. Management was primarily focused on the effectiveness of the processes carried out in the organisation.

In the last decade, the importance of product development and innovation has increased considerably. Technological updates and design renewal is important to sustain market share and a competitive position.

The challenge of innovation has, therefore, increased considerably, with limited resources and time pressure as new challenges. Markets have short-term demands for new products while the organisation, at the same time, needs to undertake long-term technology research and development. Strategic alliances are becoming a necessity in order to increase capacity in development and to give room for exploration. It is difficult to manage this development where the environment is so demanding. According to D'Aveni (1994), companies have to be innovative and constantly adapting because firms find themselves in hyper competition. Hyper competition is a competitive situation where the key success factor for companies is the ability to constantly develop new products, processes or services in a hyper active environment. Thus, companies cannot count on a sustainable competitive advantage, but must continuously innovate and grow.

The survival of companies depends on how well they adapt to changes in the environment. The competitive environment is about providing the most valuable and attractive product offers to potential customer segments at a given time (Kotler 2000; Sanches 2000). In general, the dynamics of today's business environment have increased the need for making innovation a prime objective for organisations.

In summary, in order to be successful with a BM today, one must excel in many things and, at the same time, be able to make changes toward the ever changing environment of customers and network relations. Companies need a strategy to navigate their BMs and the NEWGIBM cases have shown how catastrophic it can turn out to be if the unexpected has not been planned for. The researchers have also learned, from studying the cases, how companies can find their new business plans delayed if they do not have a strategy for their network relations as well as many other factors which affect the every day life of a company.

To start with, an explanation of what strategy is will be provided: where did it come from and what is the state of the art today?

Background: Strategy and Strategic Decisions

What does strategy mean, actually? This is a difficult question to answer, as the notion of strategy is used in so many contexts that it seems to mean everything. When this happens, one may wonder whether strategy means anything at all. A quick look at the applications of the word 'strategy' reveals that it has become one of those words which has even made its way into the everyday language. Today, people can claim to have a strategy for getting a job or a date! More seriously, strategy can be applied to the firm as a whole (corporate strategy), to parts of the firm (as in R&D strategy) and even to specific, cross functional or single functional activities within the firm (quality strategy, CIM strategy, etc.). In other words, strategy no longer has one area of application and means so many things that its concrete meaning has become blurred. There are so many different concepts at play in the strategy world today that one must wonder how all these concepts are related to each other and what all these concepts mean. The authors will attempt to provide some answers in this chapter and the starting point will be a seminal definition of strategy.

Strategy – A Seminal Definition

In his 1970 book, Russel L. Ackoff discusses the characteristics of strategy (Ackoff 1970). He comes up with three characteristics that define the concept of strategy:

* Strategy deals with concerns that are central to the livelihood and survival of the entire corporation, and usually involves a large portion of the organisation's resources and network resources
* Strategy represents new activities or areas of concern, and typically addresses issues that are unusual for the organisation rather than issues that lend themselves to routine decision-making
* Strategy has repercussions for the way other, lower-level, decisions in the organisation are made

Although these characteristics certainly may be open for interpretation, there is still an almost universal agreement on what they mean in the strategy literature today. Robert Grant, for instance, defines almost similar characteristics of strategy in his 1995 book (Grant 1995). This is not to say that there is universal agreement on the process and content of strategic management. Here something happened in the 1980s. With minor outbursts in earlier times, the 1980s saw growing differences of opinion between different authors regarding strategy. The authors of this chapter shall focus on two debates - one related to the process of strategy and one related to the content of strategy.

The Process of Strategy: From Planning to Crafting

More than any other two persons, H. Igor Ansoff and Henry Mintzberg, have been in the center of the debate on the process of strategy - mainly because their debate has become personal and rather fierce.

Back in the 1960s and prior to that, matters were relatively simple. Strategy was synonymous with strategic planning based on the SWOT notion. In other words, strategy was a plan to link opportunities and threats with strengths and weaknesses (see Andrews 1971). H. Igor Ansoff was one of many to contribute to this understanding (Ansoff 1965), with his work on strategic analysis of external and internal factors relevant to strategy. In the words of Mintzberg:

> '... So-called strategic planning arrived on the scene in the mid 1960s with a vengeance, boosted by the popularity of Igor Ansoff's book Corporate Strategy...', (Mintzberg 1994, p. 12).

Although there were many other researchers contributing to the strategic planning notion, Mintzberg is right in his basic point in the 1960s. Strategy became a popular concept among researchers and managers, and Ansoff was one of those to make that happen.

Throughout the 1980s and 1990s, however, Henry Mintzberg, and others, has been the devil's advocate in strategy theory claiming that the planning notion is theoretically unsound, i.e. human beings are not capable of understanding, analysing or comprehending the complex task of strategic decisions. Hence, strategies should be formulated as intuitive, synthesis processes. Again in the words of Mintzberg:

> *'...Strategic planning, as it has been practiced, has really been strategic programming, the articulation and elaboration of strategies, or visions, that already exist. When companies understand the difference between planning and strategic thinking, they can go back to what the strategy-making process should be: capturing what the manager learns from all sources ... and then synthesizing that learning into a vision of the direction that the business should pursue'..., (Mintzberg 1994, p. 107).*

It is to be noted that very little has been said about what the content of strategic decisions is supposed to be. The focus is mainly on how strategic decisions should be made. Nonetheless, Mintzberg's argument is that planners should not create strategies - that is the task of the entrepreneurial manager - but planners can supply data, help managers to think strategically, and later, program their visions into plans.

But what are the counter arguments? For one, Ansoff has made quite a few arguments in favor of strategic planning - both in 1990 and in 1994 (Ansoff 1990, 1994). It appears that Ansoff's arguments, apart from those personally insulting Mintzberg, fall into two categories: 1) Mintzberg's arguments are not logical and factually wrong and 2) the practice of strategic planning has lead to much more elaborate versions of strategic planning than proposed in the 1960s. Regarding the former argument, this is based on the fact that, in his papers from the 1980s, Mintzberg builds his arguments on the work of just one author. In one paper, he only uses the work of Kenneth Andrews and hence, Mintzberg fails to 'prove' much. From a logical standpoint, this may be true - even though Mintzberg's 1994 book takes into account quite a few of the authors related to the strategic planning school (Mintzberg 1994). Still, proof or not, this argument fails to even consider the content of Mintzberg's work.

This leads to the second counter argument. In Ansoff's work, as well as in that of many others, planning has been developed into a highly complex art compared to the normative proposals of the 1960s (see Ansoff & McDonnel 1990 or others). In Ansoff's own words:

> *'... In practical reality, making strategic planning flexible has been a central theme in the transmutation of strategic planning into its present variety...' (Ansoff 1994, p. 31).*

As shall be seen, this is basically true and hence it seems to be rather unfair to attack the work of the 1960s for not having developed. However, most authors contributing with new work on strategic planning still have the same basic assumptions, most notably about whether or not planning strategies are at all possible, as did Ansoff and others in the 1960s. In other words, most new approaches have failed to question their own basic assumptions, which is a natural consequence of being located firmly within one perception or paradigm of strategy. However, this does very little to convince Mintzberg and others who share different assumptions regarding the strategy process.

The Theoretical Foundation for a New View of Strategy

As already mentioned, the authors are aware of many relevant contributions to the disciplines related to strategic management. Strategy is concerned with external aspects, product-market strategy, as well as with internal factors, resources or competence-based strategy. And, as Jay Barney has observed (Barney 1995), the history of strategic management theories can be perceived as an attempt to complete the SWOT concept (normally attributed to Andrews, 1960) and move beyond strengths, weaknesses, opportunities and threats, in order to understand the competitive advantage of companies. For a discussion and review of the history of strategic management theories see Drejer (2002).

Value for the Customer as Competitive Advantage

To start externally, Michael E. Porter has been one of the major contributors to the idea that the competitive advantage of companies stems from the creation of value for the customer. According to Michael Porter, competitiveness is what the customer is willing to pay for (Porter 1985). Thus, a first attempt to define competitiveness could be to take a look at value to the customer derived from the products and services of the firm (Porter 1980, 1985). This points towards the idea that a firm can achieve a competitive advantage based on several kinds of positional advantage.

To Porter and his followers, strategy is merely a matter of creating a unique position relative to the customers of the company. It is possible to argue that this is a reactive approach to strategic management (Hollensen 2003) since it involves adapting the resources or competencies of the firm to the market conditions and competitive posture. Strategic considerations become outside-in and positioning the only key to a competitive advantage.

Since the 1980s, many have questioned parts of Porter's thinking and tried to complement his work in different ways. James Brian Quinn, in his 1992 book, quotes Ted Levitt's famous paraphrase that notes that

> '...millions of quarter-inch drill bits are sold not because people want quarter-inch drill bits but because they want quarter-inch holes. People don't buy products; they buy the expectation of future benefits...' (Levitt 1969, quoted in Quinn 1992, p. 175).

Consequently, a firm may very well offer a given (physical) product, but the customer, who defines the competitiveness of the company by his actions, may perceive the value of the product very differently from the company.

Business Development as Competitive Advantage

It is not enough to look at the customer and his or her current needs and wants. Position needs to be supplemented with other views of competitive advantage and forms of competition (Drejer 1996), with different time-horizons and characteristics in order to cope with the hyper-competitive dynamics of the new competitive landscape (Hamel & Prahalad 1994). So what are those 'other forms of competition'? Hamel & Prahalad have been among the few to discuss this explicitly, in some detail (Hamel & Prahalad, 1994) and they proposed that competition really has three sides to it: product-market competition, competition to foreshorten migration paths and competition on industry leadership, (Hamel & Prahalad 1994). The two 'new' kinds of competition are:

- **Competition for industry foresight and intellectual leadership:** This is competition to gain a deeper understanding than others on the factors (be they technology, demography, lifestyle or others) that could be used to transform current product-market competition by creating, for instance, new opportunities, new products, new ways of viewing customer needs, etc. This kind of competition has a very long time-horizon compared to the others.

- **Competition to foreshorten migration paths:** In between the battle for intellectual leadership and the battle for market share is the competition to influence the direction of industry development, i.e. a race to accumulate the necessary competencies, to test and prove as viable alternative product and service concepts, to attract coalition partners with critical complementary resources, and many other things. This kind of competition has a middle time-horizon compared to the others, leaving 'Product-Market Strategy' as the competition with the lowest time-horizon.

In short, Hamel & Prahalad's work offers an expanded view on competition, where the well-known 'competition for position' is supplemented with two other forms of competition with longer time-horizons and different contents. Nonetheless, Hamel and Prahalad (and Porter) fail to explain where the businesses of the future – the question marks – are to be developed and implemented. Hamel and Prahalad are not very operational when it comes to the how of working with the two supplementary forms of competition in a company. Here, Henry Mintzberg has been instrumental in attempting to supplement the work on 'positional advantage' (e.g. Mintzberg 1988) by emphasising business development as a cross-functional activity consisting of 'market development' via 'product development' to 'competence development' (Mintzberg 1988; Mintzberg et al. 2002). To Mintzberg, business development is the activity responsible for generating new question mark businesses into the Boston Matrix and for transforming these question marks into star businesses over time. This is an aspect of strategy that is not covered by the work on 'positional advantage' at all (Mintzberg 1988). Of course, Mintzberg is not the only one to have concluded this. Similar conclusions have been drawn in some of the contributions to the field of strategic management of technology, where some contributors speak of integrating technology development with business development (e.g. Bhalla 1987; Gaynor 1991; Dussage et al. 1991). This is also the case in the field of innovation management where the notion of innovation = invention + commercialising (Roberts 1981) where competence development and organisational change (Drejer 2001) is seen.

The latest trend in strategy research emphasises the BM concept in strategy even more. Emphasis is on businesses rather than products and on competition on BMs (Hamel 2000). When competition is on BMs, the concept of innovation also comes into focus (Drejer 2007), as innovation can be understood as the way new BMs are developed and evolve. When innovation is important, the network organisations are crucial to management in the future (Chesbrough 2006).

Dynamic Capabilities as Competitive Advantage

For the last 10-15 years, many have come to realise (again) that the resources of the company can also be a source of competitive advantage and enable the company to develop new products and services, and even businesses for the market, rather than consider the organisation as a necessary evil (Eriksen & Mikkelsen 1993; Wernerfelt 1984; Barney 1991). This is mirrored in the vast attention devoted to concepts such as 'the learning organisation' (Argyris & Schön 1978, 1998; Duncan & Weiss 1979; Watkins & Marsick 1995; Nevis et al. 1995), 'redesign of business processes' (Hammer & Champy 1993; Champy 1995; Harrington 1991), 'human resources' (Tichy et al. 1982; Garvin 1993; Senge 1990), 'mission statements' (Cambell & Tadaway 1990; Cambell & Yeung 1991; Andersen 1987) and many other new management fads. For an excellent review of the development of new management fads and what they seem to have in common, see Sharpiro (1996) or Drejer (2002) for the strategic version. Nonetheless, the last 10-15 years have provided a supplementary understanding of how companies can create a competitive advantage. This has happened via the resource-based view of strategy (e.g. Wernerfelt 1984; Eriksen 1993a; Grant 1991), 'critical capabilities' (e.g. Hayes et al. 1996; Teece et al. 1997; Eisenhardt & Martin 2000) and, not in the least, 'core competencies' (e.g. Bogner & Thomas 1996; Hamel & Prahalad 1994; Drejer 2002).

Normally, the concept of 'core competencies' is attributed to Prahalad and Hamel as seen in Prahalad & Hamel (1990) and in Prahalad (1993), so why not start there? In 1990, Hamel and Prahalad defined 'core competencies' as:

> '...the collective learning in the organisation, especially how to co-ordinate diverse production skills and integrate multiple streams of technologies'..., (Prahalad & Hamel 1990, p. 82).

Furthermore, Prahalad (1993, p. 45) elaborates on this as:

Competence = (Technology • Governance Process • Collective Learning)

Regarding 'core competence', Hamel and Prahalad say it must be subject to three tests:

- Is it a significant source of competitive advantage?
- Does it transcend a single business?
- Is it hard for competitors to imitate?

This points towards the main contribution of Hamel and Prahalad and their followers, which is the definition of competencies as a part of strategic management. From a strategic view-point, 'core' competencies are those competencies that provide the firm with a competitive advantage via the execution of the competence. Examples of this are the way in which Danfoss manages its worldwide innovation setup, KMD's relationship to the Danish local councils and its many network partners and DixiTel's know-how about the mobile phone industry and such. Core competencies have been built up over time and are not easily imitated. For instance, it would be very difficult to gain access to the knowledge of Danfoss managers on innovation and, not to mention, to imitate Danfoss' credibility with the customers. Core competencies are distinct from competencies that are not 'core'. Such other competencies may be either enabling or supplemental competencies. 'Supplemental' competencies are those that add value to the 'core' competencies but could be imitated. Examples include the distribution channels of DixiTel or KMD's software for gathering data on energy consumption. 'Enabling' competencies are necessary but not sufficient in themselves to distinguish a firm competitively. For instance, quality management skills and systems are increasingly the price for entering the competitive game rather than the road to superiority. Evidently, some core competencies will, over time, slip and become 'enabling' competencies if nothing is done to preserve the 'core' competence and make it impossible to imitate. This can happen if the core competence contains proprietary knowledge (unavailable from public sources) or is developed constantly beyond the competitor's competencies.

It is important to note that the contributions on core competencies emphasise organisational learning as an important part of strategic management (e.g. Wick & Leon 1993; Pedler et al. 1991; Nonaka & Takeuchi 1995). This implies that knowledge and skills are important elements of the competencies in a company (e.g. Edivinson & Malone 1997; Stewart 1997; Davenport & Prusak 1997; Winter 1987). At the same time, this implies that competencies need to be viewed as dynamic rather than stable entities (Eisenhardt & Martin 2002; Teece et al. 1997) that should be subject to continuous competence development over time (Drejer 2002). Still, there is something missing from the seminal definitions. The work of Hamel and Prahalad fails to explain how core competencies are linked to other aspects internally within the firm. For instance, what about knowledge and knowledge workers? Or what about technology? Furthermore, Hamel and Prahalad's definition of competencies is clearly a functional definition in the sense that it clearly emphasises what the competence does: is it important to our customers? – rather than focus on how this is done. This points towards the need for a structural definition of competencies. This is also much needed in order to find out how different kinds of competencies look, how they change over time and how to describe, analyse and develop competencies in networks, no less relevant, in light of the discussion of business development and cultural clashes. One reason for the emphasis on developing competencies ties in with another part of the discussion, namely innovation. For instance, Leonard-Barton quoted Gary Hamel for saying that competence provides a gateway to new opportunities that the organisation must exploit in order to survive (Leonard-Barton 1995). In doing so, the knowledge and competencies that the firm is based upon must also change as:

> '... even seemingly minor innovations that alter the architecture of a product can undermine the usefulness of deeply embedded knowledge ...', (Leonard-Barton 1995, p. 17).

Thus, innovation and BM development are key-factors in rendering current core competencies obsolete. This phenomenon is called 'core rigidities' by Leonard-Barton. Drejer and Riis argue that it is necessary to define both core competencies (the strategically important competencies at present) and focus competencies (the strategically important competencies in the future) in order to initiate the necessary competence development in due time (Drejer & Riis 2000) and to develop competencies in networks (Drejer 2004).

Definition of Strategy and Business Development

What, then, is 'Strategic Innovation'? As early as in the 1950s, Peter Drucker distinguished between 'doing the right things' and 'doing things right' (Drucker 1958). 'Strategic management' can be reformulated to: First, to market the right products/services in the right markets and, second, to develop, produce, and distribute the products/services in the right way. It is intuitively clear that organisations and organisational networks need to focus on both issues, in the long run, while maintaining, at the same time, a dual focus on business development and operational effectiveness. The foundation for the present chapter on 'strategic innovation' is that the authors think of 'strategy' as:

- Change of the position of the organisation and/or an organisational network in the marketplace at the same time as exploiting the current position.

- The environment consists of present, as well as potential customers or chain of customers and of a large number of different players. It is the entire environment of the organisation or the organisational networks that needs to be taken into account in strategic management.
- The organisation or organisational network itself should be seen as a holistic entity consisting of business and resources. This means that the strengths and weaknesses of the organisation or organisational network should be described in terms of 'bundles of resources' or 'competencies' rather than as 'functional units'.

In consequence, the potential of the existing resources in a network to create value in marketplaces different from the current one (W. Chan Kim and Renée Mauborgne 2005; Chesbrough 2007), while still creating value in the current situation, becomes an important consideration in 'strategic management'. One may speak of a 'competence readiness' that the company possesses and is able to apply by reorienting its business foundation towards new marketplaces, i.e. 'strategic innovation'.

'Strategic innovation' can also be defined as:

> *'business modelling', the ability to create and revitalise the business idea and model of the company by changing both the market of the organisation/organisational network and the competencies and business system of the organisation/organisational network. In this way, 'strategic innovation' is concerned with developing the entire organisation/ organisational network…'*

A Conclusion on the Proposed View of 'Strategy' and 'Business Models'

A 'business model' is the combination of a business idea, a business organisation, and a business system (strategy) and is the super concept at play. When discussing 'strategic innovation' the purpose of innovation is to develop new BMs.

Something not mentioned in the 'framework for analysing strategy innovation' is the why question. The answer to this question answers the strategic assumptions of the BM. Is the business based mainly on a group of customers? Is it on certain services or maybe on certain competencies? And why is it so? This determines a lot of things about the business besides the other issues of the BM. 'Why' goes beyond the traditional economical explanations for doing business. It concerns a higher motivation: making the world a better place to live in, looking after the environment, spreading social innovation etc. More companies will seek new answers to the question of why they are doing business and they will concern themselves with the need to deliver experiences and to provide more reason in our future society. As organisations do this, they will find that working in networks and defining businesses in terms of networks will prove to be a very viable approach.

References

- Ackoff, 1970.

- Andrews, 1971

- Ansoff, 1965 Ansoff, 1990, 1994

- Barney, 1995

- Chesbrough, 2006.

- D'Aveni 1994

- Drejer, 2007 Drejer, 1996, Drejer, 2001

- Drucker, 1958

- Grant, 1995.

- Hamel, 2000

- Hollensen, 2003

- Leonard-Barton, 1995, p. 17

- Levitt, 1969, quoted in Quinn, 1992, p. 175

- Mintzberg, 1994, p. 12 Mintzberg, 1988; Mintzberg et al., 2002

- Porter, 1980, 1985

- Prahalad, 1993

- Roberts, 1981

- Sharpiro, 1996

- Zeleny, 2007

- Ansoff & McDonnel, 1990

- Argyris & Schön, 1978, 1998; Duncan & Weiss, 1979; Watkins & Marsick, 1995; Nevis et al., 1995

- Bolwijn and Kumpe 1994.

- Cambell & Tadaway, 1990; Cambell & Yeung, 1991; Andersen, 1987

- Drejer & Riis, 2000

- Eisenhardt & Martin, 2002; Teece et al., 1997

- Eriksen & Mikkelsen, 1993; Wernerfelt, 1984; Barney, 1991

- Hammer & Champy, 1993; Champy, 1995; Harrington, 1991

- Kotler, 2000, Sanches, 2000.

- Russel L. Ackoff 1970

- Prahalad & Hamel, 1990 Hamel & Prahalad, 1994

- Tichy et al., 1982; Garvin, 1993; Senge, 1990

- W. Chan Kim og Renée Mauborgne 2005. Chesbrough 2007

- e.g. Bhalla, 1987; Gaynor, 1991; Dussage et al., 1991

- e.g. Bogner & Thomas, 1996; Hamel & Prahalad, 1994; Drejer, 2002

- e.g. Edivinson & Malone, 1997; Stewart, 1997; Davenport & Prusak, 1997; Winter, 1987

- e.g. Hayes et al., 1996, Teece et al., 1997; Eisenhardt & Martin, 2000

- e.g. Wernerfelt, 1984; Eriksen, 1993a; Grant, 1991

- e.g. Wick & Leon, 1993; Pedler et al., 1991; Nonaka & Takeuchi, 1995

Chapter III.6

Successful Implementation of Global BM Innovation

Yariv Taran

&

Peter Lindgren

Introduction

The business model concept continues to evolve and embrace new perceptions, challenges and opportunities. The case studies that were analysed in the previous chapters have changed the views and understanding of the researchers concerning the functions of the BM concept. It is the researchers' belief that the BM concept serves as more than the translation of a company's strategy into a visual blueprint. It should also be considered as a managerial tool for new strategy formulation.

Although the point of departure for the NEWGIBM project (New Global ICT Based Business Model) was the BM concept applied to a single company, the researchers noticed that the global perspective is strongly related, not only to the ICT basis of the firm, but also to the 'power' of a company's network ties or 'relation capital'.

For that reason, the researchers believe that the BM concept should be analysed from a network perspective. It is not enough to visualise who the network partners of the firm are, as suggested by the single company BM framework, but also to draw a framework of how to form the network-based BM and how to innovate it, in addition to recognising how to lead the network successfully.

In view of that, this chapter will attempt to provide a new framework for business model innovation, which could, and should, be analysed, both from a single company viewpoint, as well as from a network-based perspective.

Business Model vs. Strategy

Before continuing, it is essential to distinguish between two close, but still somewhat different concepts: 'business model' and 'strategy'. Both concepts are used frequently, both by researchers as well as by company leaders – but is it truly possible to define the differences between the two concepts? According to Magretta (2002), both concepts are among the most sloppily used terms in businesses, 'they are often stretched to mean everything – and end up meaning nothing'.

Seddon et al. 2004 have tackled this issue, using an in-depth research approach to clarify the differences between the two terms. As part of their study, they found the following two answers to the question mentioned above:

* 'Business models describe, as a system, how the pieces of a business fit together. But they don't factor in one critical dimension of performance: competition...

Figure 1: Possible Overlap Between 'Strategy' and 'Business Model' Concepts
(Seddon et al. 2004)

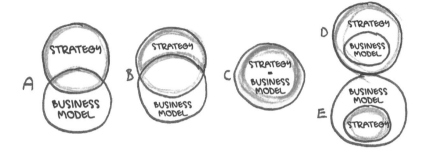

Dealing with that reality is strategy's job.... A competitive strategy explains how you will do better than your rivals' (Magretta, 2002).

- Strategy seems more concerned with competitive positioning, whereas BMs are more concerned with the 'core logic' (Linder and Cantrell, 2000) that enables firms to create value for their customers and owners.

On the basis of this, Seddon et al. stated that their research has led them to believe that the business model should be viewed and defined as an abstract representation of one aspect of a firm's strategy. This means that, unlike strategy, business models do not take into consideration a firm's competitive positioning. Furthermore, BMs are inward oriented, focusing more on the operational effectiveness of the firm and the 'activity-system' aspect of how the firm creates economic value, while the strategy is outward oriented and focuses more on the competitive positioning of the firm.

From the figure above, it can be concluded that business models are the translation of a company's strategy into a visual blueprint (sketch 'D' in Figure 1).

Still, unlike Seddon et al. (2004), the present research has led to the belief that business model innovation is generated directly from a new strategy formulation. Thus, a single strategy can lead to only one business model design and not to two or three BM designs as presented in the figure above. This is, of course, related to the classification of when a BM can be defined as new. Nonetheless, the findings of Seddon et al. (2004) have provided the NEWGIBM researchers with a starting

Figure 2: Relationship Between the Two Concepts: 'Strategy' and 'Business Model'
(Seddon et al. 2004)

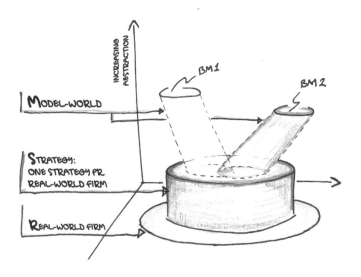

point as to how to understand the differences between the two concepts, as well as how to appreciate the relation that exists between the two concepts.

Roadmap to Global BM Innovation:

Without a doubt, the theoretical background and the case studies presented, thus far, have contained very extensive amounts of data. Yet, what is really known about the underlying principles of implementing a business model innovation? It seems that all of the business model research 'gurus' have forgotten to mention what is the true purpose of clarifying the BM concept in the first place: is it a managerial tool – or simply a 'picture' to hang on the CEO's office wall?

Since BMs are the translation of a company's strategy into a visual blueprint, they serve as a fundamental managerial tool that can assist with new strategy formulations (both for the single company as well as for the overall network). For that reason, managers should utilise their BMs as part of their strategy formulation process in order to assist them with making more rational future strategic planning. This can be accomplished by means of:

- Discrimination and prioritiation between options
- Evaluation of mergers, acquisitions, or joint ventures
- Identification of the resources needed to implement a certain initiative
- Monitoring business expansion and change over time, cost efficiencies, fund raising support, etc.

As mentioned in the beginning of the book, the greatest challenge to the NEW-GIBM research project was to link together the rising gap between BM theories and the practical implementation of them. A typical question asked by some leaders within the case studies was: 'Leaving theory aside - what do I need to do "Monday morning" in order to innovate my business model?' The figure above, and the next part of this chapter, is aimed towards answering exactly that question.

The road map of global business model innovation presented here, is a result of both the individual research of the NEWGIBM group, as well as network relations with top researchers in this field (Chesbrough, Osterwalder and Tucci). The

Figure 3: The Four Step Roadmap to Global BM Innovation.

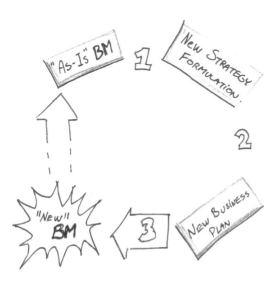

purpose of this road map is to illustrate a practical implementation process of global business model innovation in a linear and systematic manner. First, each step should be analysed from a single company viewpoint and, later on, from a network perspective.

Step 1: 'AS-IS' BM Description:

A business model serves as a building platform that represents a company's operational and physical manifestation. Thus, the challenge for a business model designer, before innovating the business model, is first, to identify the key elements and the key relationships that describe a company's 'AS-IS' business model before innovating it.

The present chapter separates the 'AS-IS' BM process description into two phases. In the first phase, each company visualises its own BM blueprint. In the second phase, the entire network is challenged to visualise their network-based BM or the network 'AS-IS' interface platform

Table 1: Describing the 'AS-IS' BM

Pillar	Business Model Building Blocks	Step 1: Single-Level	Step 2: Network-Level
Product/services/ processes	Value Proposition	What 'value proposition' do **I** (the individual company) offer? (Products, services, processes)	What 'value proposition' do **we** (the firms in the network) offer? (Products, services, processes)
Customer Interface (physical, digital, virtual)	Target Customer	Who do **I** create value for? (To whom? To which customer segments?)	Who do **we** create value for? (To whom? To which customer segments?)
	Distribution Channel	How do **I** deliver the products/services to **my** customers at the appropriate cost?	How do **we** deliver the products/services to **our** customers at the appropriate cost?
	Relationship	What kind of relations do **I** establish with **my** various customer segments?	What kind of relations do **we** establish with **our** different customer segments?

Pillar	Business Model Building Blocks	Step 1: Single-Level	Step 2: Network-Level
Infrastructure Management	Value Chain Configuration (Internal)	How do **I** create value? (Activities and resources)	How do **we** create values? (Activities and resources)
	Competencies	What are **my** sources of competencies?	What are **our** sources of competencies?
	Network	Who are **my** netw. partners?	(not relevant)
Financial and Non-financial Aspects	Turnover and Cost Structure	How often do **I** compare the financial results to the budget? (Turnover and Cost Structure)	How often do **we** compare the network financial results to the budget? (Turnover and Cost Structure)
	Revenue Model	How do **I** make money in this business? (Revenue model)	How do **we** make money in this business/network? (Revenue model)

The table above is based on core questions and each core question can be further specified into numerous sub-questions. The purpose of this chapter, however, is 'pointing in the right direction' towards business model innovation rather than attempting to provide a holistic framework to it.

Step 2: New Strategy Formulation

Once the BM of the firm (and, later on, the BM of the network) has been visualised, the company (or network) can move into the second phase by innovating its strategy, which, to a certain extent, has been visualised in step one.

As part of the new strategy formulation process, it is essential for leaders to understand that strategy innovation must contain an aspiration (**what** to innovate?), a practical intention (**how** to innovate successfully?), and an explicit purpose (**to whom**?).

Figure 4: Distribution of Innovation Capability
(source – Tidd at al. 2001)

An interesting question arises when researchers and leaders try to address the Who question, namely – who should the company innovate with? Should the company innovate single handedly or in networks?

Although the figure above perceives innovation purely from a technological perspective, it can still provide researchers with an interesting conclusion to the question mentioned above. The decision of companies to open their BM to network-based innovation is strongly related to their readiness to change, as well as their core competence capabilities. Firms labelled as 'type one' are very static in nature. Their readiness for change is small which, in consequence, is also the case concerning the innovation potential of the company. Firms labelled as 'type four' are highly capable of innovating successfully single-handedly. However, firms in the categories 'type two' and 'type three', where most companies are situated today, are faced with a great necessity of opening their BM for external networking innovation support, since the competences that they hold in-house are not sufficient for sustainable (continual) 'single-handed' innovation.

Figure 5: Framework for Analysing Strategy Innovation

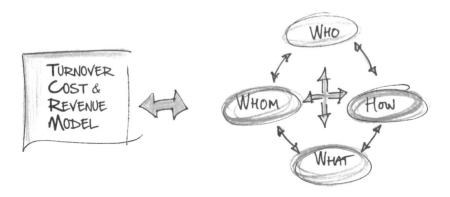

As part of the NEWGIBM project, an analytical model (figure below) was developed.

The interaction among the four main categories of questions results in six important analytical areas. The purpose of the table below is to assist the user in considering possibilities of future innovation. The following interactions are not ranked after importance. The ranking will depend entirely on the specific case. As in phase one, the interaction between the groups into individual company and network based levels has been separated.

Table 2: Interaction Between the Groups

	Step 1: Individual Company Level	Step 2: Network Level
Who -> What	Who are the stakeholders that **I** (the individual firm) plan to work with in order to develop the new product, process or service chosen? (Here, **I** have an idea (what), but am looking for somebody (who) to work with in order to implement it). What is it that I want to innovate with my chosen stakeholders group? (Here **I** use stakeholder groups to generate an idea). What is it that I want to innovate? And who do I want to innovate it with? (Here **I** am very much in the 'dark' – and try to outsource both the idea as well as outsource the network partners).	Does the network of participants have all the necessary resources (skills, competencies, capabilities etc.) needed to develop the product, process or service? If not, who should **we** (the network firms) include in the network? Does the network have a clear understanding of the value-proposition that **we** intend to present to the customer?
Who -> To Whom	Assuming that **I** have already chosen **my** customer (whom), then who are the stakeholders **I** need to work with in order to best provide for **my** chosen customers? Assuming that **I** do not yet have a target group, who are the stakeholders that **I** need to work with in order to define **my** new/ chosen customer group?	Do the participants in the network have the necessary knowledge about the customer/market that **we** need to make a successful entry of the product, process or service? If not, how can **we** get hold of the knowledge? What are the competitive products? What is the market size?
Who -> How	Assuming that **I** have the answer to 'how', who are the stakeholders needed to assist **me** in launching the distribution/ innovation of the new product, process or service? Assuming that **I** do not know how to distribute/ innovate, who are the stakeholders needed to advise **me** as to how to distribute/innovate? Unlike the conjunction between 'what' and 'who', where the company concentrates on what it wants to innovate, **I** am more concentrated on the level of innovation in terms of radical vs. incremental.	Do the participants in the network have all the necessary resources (skills, competencies, capabilities etc.) that **we** need to distribute the product, process or service in the manor that **we** want to?

	Step 1: Individual Company Level	Step 2: Network Level
What -> To Whom	Assuming that **I** have the answer to whom (**my** customer), what new products, processes or services are needed for **my** chosen customer target group(s)? Assuming that **I** do not have a target group yet, to whom do **I** want to innovate, and what?	Does the product (distributed by the network-partners) live up to the needs and desires of the customer?
What -> How	Is the chosen method of distribution/ innovation the appropriate choice for this particular product, process or service?	Is the chosen network of distribution/ innovation the appropriate choice for this particular product, process or service?
To Whom -> How	Is the chosen means of distribution/ innovation the appropriate choice for this particular customer/market?	Is the chosen network of distribution/ innovation the appropriate choice for this particular customer/market? Product development – Changes and ways of change through the network process.
Revenue - model	Which type of revenues model do **I** have? Is the purpose of the revenue model cost efficiency, knowledge gaining or increasing revenue? Is the purpose of the revenue model long-term or short-term oriented?	Which types of revenues do the network partners have before entering the network formation? Did the network partners change their revenues perceptions after entering the network? Both financially and non-financially? Is the purpose of the network formation cost efficiency, knowledge gaining or to increase revenues? Is the purpose of the network formation long-term or short-term revenue model oriented?

Step 3: New Business Plan

Once the new strategy has been formulated, the company (or network) can move into the third phase of designing a new business plan. This does, however, raise the question of where to draw the borderline between a company's business model and a company's business plan?

Unlike the BM, the business plan is a very short description of a company's strategy objectives. It is not meant to tell the users the whole story. Therefore, the business plan does not discuss in details any specific operational processes, organisational structures, or financial forecasts. Furthermore, the business plan does not explain which sets of building blocks are being used in the process, nor their relation to each other.

The purpose of the business plan is, exclusively, to allow a preliminary visualisation of the new business or network and its success rate (SWOT analysis) based on the chosen strategy. Additionally, many small and medium sized companies use a business plan to create interest in order to attract investors or new network partners. Therefore, the business plan serves both inward and outward intentions.

Step 4: New BM Drawing

By comparing phase four with phase one, leaders could easily identify the extent of their BM change.

Table 3: Describing the 'New' BM

Pillars	Business Model Building Blocks	Step 1: Single-Level	Step 2: Network-Level
Product/services/ processes	Value Proposition	What new 'value proposition' do **I** (the individual firm) offer? (Products, services, processes)	What new 'value proposition' do **we** (the firm and its network partners) offer? (Products, services, processes)

Pillars	Business Model Building Blocks	Step 1: Single-Level	Step 2: Network-Level
Customer Interface (physical, digital, virtual)	Target Customer	Who do **I** create value for? (To whom? To which new customer segments?)	Who do **we** create value for? (To whom? To which new customer segments?)
	Distribution Channel	How do **I** deliver the new products/ services to my customers at the appropriate cost?	How do **we** deliver the new products/services to our customers at the appropriate cost?
	Relationship	What new kind of relations do **I** establish with my different customer segments?	What new kind of relations do **we** establish with our different customer segments?
Infrastructure Management	Value Chain Configuration (Internal)	How do **I** create value? (New value chain configuration – activities and resources)	How do **we** create value? (New value chain configuration – activities and resources)
	Competencies	What is **my** new source of competencies?	What is **our** new source of competencies?
	Network	Who are **my** new network partners?	Who are **our** new network partners?
Financial and Non- financial Aspects	Turnover and Cost Structure	New turnover and cost structure	New turnover and cost structure
	Revenue Model	How do **I** make more money in this business? (Revenue model)	How do **we** make more money in this business/ network? (Revenue model)

As will be seen in the next section, in some of the case studies, the network partners were convinced that their BM change should be considered as radical but once the change was visualised and compared with the original BM structure, the actual change could be described as relatively small.

Case Studies Examples for BM Innovation

As part of the NEWGIBM case studies, it has been clearly identified that not all of the participants in the network were interested in the same results, or focused on the same revenue model. Some of the participants, for example, valued knowledge generation more than the actual financial benefits. Still, although the desires of the different participants may have been very diverse, they all shared a common understanding that it was within their interest to stay in the network, even if it meant that they needed to adjust their own BM formation to meet the BMs of the network partners. In this way they could benefit from the overall core competences and skills of the network, whether the expected benefits were financial/ non-financial, short-term or long-term oriented.

Another interesting point that arose from the NEWGIBM case studies was that the more global the network becomes - the more it is essential for the network partners to learn to appreciate the importance of ICT usage because of the fact that most of the interaction between the partners is based on virtual communication.

The findings of the present case studies have also shown that the newness of a network-based business model is strongly related to the structure of that model as a whole, and not to a specific company's 'openness'[1] of their business model change. Moreover, although the changes in individual BM formation can be fairly incremental when entering the network, they can still, in some cases, lead to a relatively radically new product or service launched by the entire network (KMD Case).

Business model innovation may either result from an equal partnership (joint venture), or take place within a loose, more diversified network (Smart Harbour). In either case, some network partners will be forced to change their individual business logic more radically than others. In addition, the development of a network-based business model may result in a radical change of customer focus, since it involves new technologies, new value propositions, new value chains, new network formations, and new markets (potentially the Smart House).

---

1   Chesbrough, H (2005)

Figure 6: Five + 1 Phases of Evolution Within the BM Literature

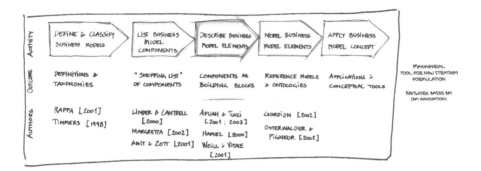

Furthermore, in some cases (Smart House) the new network formation ties were not necessarily characterised by their industrial homogeneity or relations with various governmental or consultant agencies (vertical, horizontal or diagonal networks). They were characterised instead, by the large diversification of its network partners' products and services (across industrial borders) – for the purpose of pursuing radical innovation ('blue ocean²') possibilities.

All in all, the present research project has led the authors to believe that the business model concept continues to evolve and embrace new perceptions, challenges and opportunities. The case studies that have been analysed here have changed the authors' views and their understanding of what the BM concept serves as. It is the belief of the authors that the BM concept serves as more than the mere translation of a company's strategy into a visual blueprint. It should also be considered as a managerial tool for a new strategy formulation, both for an individual company, as well as for the entire network BM.

Conclusion

The aim of this chapter was to introduce and illustrate a framework of successful implementation of global BM innovation at the level of a single company, as well as at a network level. The point of departure was the core elements of firm-level models (Osterwalder et al. 2004). The NEWGIBM case studies

---

2   Chan and Mauborgne 2005

suggested the following differences between firm-level and network-level business models:

- A network-based business model is a powerful tool for individual companies to innovate with, particularly when they recognise that the core knowledge and/or competences needed are not available in-house. This point is applicable to all phases of the innovation process from idea-generation to application (Smart House, Master Cat).
- The level of innovation is potentially high with a network-based business model, due to the large mixture of competences and ideas becoming available to the partners (Intelligent Utility, (potentially) Smart House/Smart City).
- The 'time to market' of a network-based innovation can be reduced significantly compared to single-firm innovation, either because of an immediate exploitation of existing technologies, faster market introduction, or due to higher capital liquidity utilisation capabilities (Master Cat, Intelligent Utility).

The new network-based business model has a much stronger focus on the basics of how innovation ultimately creates value and reduces costs for both customers, suppliers and others participants in the innovation process. It focuses on the creation of both value and on costs.

Although it is, at present, quite difficult to foresee which of the involved new global business models in the NEWGIBM project will eventually be successful, there is some evidence showing which cases went successfully through the innovation and concept phases. The KMD, Smart House and Smart Harbor projects, in particular, came very close to prototyping and market implementation. The Tricon project failed and one project was 'rejected' by the market. The Danfoss project is still in progress and the Isopaint and one other case came to a stand-still until the necessary resources and manpower are made available.

Further Research

The network-based business model proposed within this chapter can, and should be, tested on a large scale survey, possibly supported by more in-depth case studies. For that reason, the authors plan to carry out similar research on an international basis (within Europe and the US). It is also the aim of the authors to be able to draw more concrete conclusions regarding the global aspect of new business models.

References

- Chan, W. K. and R. Mauborgne (2005), *Blue Ocean Strategy, How to Create Uncontested Market Space and Make Competition Irrelevant,* Harvard Business School Press

- Chesbrough, H (2005), *Open Innovation The new Imperative for creating and Profitting from Technology,* Harvard Business School, Press

- Chesbrough, H (2007), *Open Business Models How to Thrive in the New Innovation Landscape,* Harvard Business School ISBN 13: 978-1-422-1-0427-9

- Linder, J. and S. Cantrell (2000), *Changing Business Models: Surfing the Landscape,* Accenture Institute for Strategic Change, Canada

- Magretta, J. (2002), *Why Business Models Matter?* Harvard Business Review, Vol. 80, No. 5, pp. 86–92

- Osterwalder, A., Y. Pigneur and L.C. Tucci (2004), *Clarifying Business Models: Origins, Present, and Future of the Concept,* Communications of AIS, No. 16, pp. 1–25

- Osterwalder, A. and Y. Pigneur (2004), *An Ontology for e-business Models,* University of Lausanne, Switzerland

- Seddon, P.,., Lewis, G.. & Shanks, G. (2004). *The Case for Viewing Business Models as Abstractions of Strategy.* Communications of the Association for Information Systems, 13, 427–442

- Tidd, J., Bessant, T. & Pavitt, K. (2001) *Managing Innovation,* 2nd edn, (Chichester: Wiley)

Chapter IV.1

Globalisation Of ICT Based Business Models: Today And In 2020

Peter Lindgren
Yariv Taran
Kristin Saughaug
&
Andreas Slavensky

Abstract

Companies today, in some industries more than other can now see the EMERG-ING ICT based GLOBAL BUSINESS MODELS based on MOBILE AND WIRELESS technology. Communication will in 2020 look different to what we know about today.

Due to heavy investment in mobile and wireless research and technology both from private and public bodies developed and diverse solutions we hardly can imagine will be realized and even common to us in 2020.

Time is now to think about which new mobile and wireless communication business model opportunities we will enter. However, the choice, development and innovation of these new technical and business models are indeed a complex matter.

The history of ICT based Communication is relatively young and the story on Mobile and Wireless Communication based business models is even younger. However the study of Mobile and Wireless communication together related to multi business models has just begun.

In 2020 the user will be placed in the middle of these technologies - in the very core. The technology will be user centric and will finally enable the user to con-tinuously fulfill his demand of new, radical, speedy development of new value possibilities. The supply of new values will up till 2020 be tenfold previous seen evolutions.

This chapter will explain firstly the history and background of ICT based com-munication – in specific Mobile and Wireless communication and its relation to the new business models, hereafter the explain how we see these new models and then finally offer a futuristic Outlook to multibusiness Models and innovation possibilities enabled by ICT in 2020 .

Keywords: ICT based business models, Business Models, Innovation, multi business models

Introduction

Mobile and wireless communication are a challenge to many innovators and es-tablished companies. Especially the old print media companies (Newspapers,book publishers e.g.) are indeed bleeding in these days due to that mobile and wireless technologies have entered the global market place. The strategic question is - should one leave the old technology and business model and join the new technology and market? Should one invest or just wait for technology to improve more?

Business leaders, when asked to explain their future company's strategy to business model in the mobile and wireless communication industry is very diverse. Some invest heavily - some just follows the development waiting for technology, standards and proved business models to come. Many are waiting for critical mass of customers and business. Others with quiet interesting running business models on behalf of the mobile and wireless communication technologies and markets (Apple, qq.com, Cisco) are often very closed or not particularly very communicative about how and where they do their business.

When they however do come up with an answer – Apple, HTC, Google, Nokia e.g. - they would most likely comment and present their new vision for new technology and business models together with their bid on future structure rather than comment on existing business. It is therefore very hard to see if they really are making a profit on mobile and communication technology.

Indeed much technology and many business models in mobile and wireless communication industry is still in the very introduction phase (Wind 1973) – and many business opportunities are still question marks (Wind 1973). However nobody can deny that we are indeed moving into a mobile and wireless communication society – and fast - but do the mobile and wireless communication based markets represent one favourable business model? Many managers at the moment do not really know and few can really imagine what the future business model with this technology will bring us and what the future business model will be about and will represents (Printed Media 2010, Times 2010, Green Tech Media 2010).

How will the moble and wireless communication technology and business models really function in 2020? Do they even have one explicit business model? And assuming they do, do we know how to commercialise and continually innovate these – to create successful businesses that will be able to grow and survive?

However lets take a look back on the history and development of the mobile and wireless industry and lets try to understand what this opportunity and industry is about Then lets have a look on what to become and where we expect to be in 2020.

The History and Evolution of the Mobile and Wireless Communication Business Models

What are the mobile and wireless communication technology and business model about. Basically today in 2010 the Mobile and Wireless communication technology are about any mobile and any wireless communication in our environment (Paavilainnen 2002). However in 2020 it will be about any human and business centric activity enable and often conducted by mobile and wireless telecommunication network.

The Mobile and wireless communication industry is not a variation on existing internet services – it is a natural extension of our mobility and living platform. Some would say we finally are given our opportunity of freedom back – because we now will be able to communicate anywhere, anytime, with anybody and with anything. Previously the technology bound us to home, offices and central points where technology allowed us to communicate. The future is opposite.

However it is not just about us getting our freedom back but it is also allowing us via our mobile and wireless devices finally to freely communicate and be mobile at the same time.

How did we reach this point. Well the journey to this stage game through several generations of wireless technology – from 1G to 4G and now towards 5G.

Table 1: From 1G to 4G and now towards 5G.
Lindgren 2010 inspired from Efraim Turban Electronic Commerce – an managerial perspective 2010

Generations of wireless communication technology	Description	Time period
1G	The first generation of wireless technology. It was an analog-based technology used exclusively for voice.	1979 - 1992
2G	The second generation based on digital radio technology and is able to accommodate voice and text messages (SMS)	

Generations of wireless communication technology	Description	Time period
2.5G	An interim technology based on cell phone protocols such as GPRS (General Packet Radio Service and CDMA2000 (Code Division Multiple Access). This generation can communicate limited graphics, such as in picture text messages (EMS)	
3G	Digital wireless technology, which support rich media such as video. 3G utilizes packet switching in the high 15 to 20 MHz range.	2001 -
4G	Provide faster display of multimedia	2006 -
5G	Real wireless world - completed WWWW: World Wide Wireless Web	-

6 billion people own a mobile phone today and we have been through various generations of cellular systems in the evolution of our mobile communication from 1st generation to 4th generation. Now almost all the service providers as well as the customers seek for availing 3G and 4G mobile and wireless services. Daily increase in the increase of mobile and wireless customers - increase the potential and reachable market. Mobile and wireless companies is launching high tech featured mobile into the market continuously.

In 2010 we are in the midst of 3G. In China, the 3G service came into existence only by February 2009. It may take time for exposing to other states, but we should also realize many other countries such as the Asian countries were using these services since last decade.

Thanks to heavy research and big investments in mobile and wireless technology we are now heading quickly towards 5G may which we expect will enter within few years. We finally reach the end of the beginning – 5 G. We have started to create the first part and a new digital communication layer on our world. 5G - Real wireless world - completed WWWW: World Wide Wireless Web will be started from 4G technologies. The evolution will be based on 4G and completed into its idea - a real wireless world. Thus, 5G will make an important difference and add more services and benefit to the world. 5G will be a more intelligent technology and will create more intelligent business models that interconnect – but even more interconnect with the entire global market and without limits.

The technological platform leaves us with some tremendous amount of new applications and opportunity all based on what we did in the past on mobile and wireless communication technologies and we can possibly innovate in the time to come.

However we have not even touched the real potential of these technologies – seen in a business model perspective – because few of us can really imagine what we really can do with this new platform. Still we are playing with the technology and the business models on a rather simple level.

It leaves us with some important choice and decisions to take – which technology and business models should we go for. Just to give an overview of some of the classes of mobile and wireless application we are given were we stand in 2010 – this nice piece of work from Efraim Turban 2010 gives a brief overview of some of the classes of Mobile and the wireless applications.

Table 2: Classes of Mobile and wireless applications
Classes of Mobile and wireless application inspired by Efraim Turban 2010

Class of Application	Examples
Mobile financial applications (B2C, B2B)	Banking, brokerage, and payments for mobile users
Mobile advertising (B2C)	Sending user-specific and location-sensitive advertisements to users
Mobile inventory management (B2C,B2B)	Location tracking of goods, boxes, troops, and people
Proactive service management (B2C,B2B)	Transmission of information related to distributing components to vendors

Class of Application	Examples
Product locating and shopping (B2C,B2B)	Locating/ordering certain items from a mobile device
Wireless reengineering (B2C,B2B)	Improvements of business services
Mobile auction or reverse auction (B2C, B2B)	Services for customers to buy or sell certain items
Mobile entertainment services (B2C, B2B)	Video-on-demand and other services to a mobile user
Mobile office (B2C,B2B)	Working from traffic jams, airport, and conferences
Mobile distance education ((B2C,B2B)	Taking a class using streaming audio and video
Wireless data center (B2C, B2B)	Information can be downloaded by mobile users/vendors
Mobile music/music-on-demand (B2C,B2B)	Downloading and playing music using a mobile device

It leaves us with some new digital attributes, opportunities and benefits which we have just begun to investigate and understand. Mobility – is based on th fact that users and items carry a cell phone or other mobile devices everywhere they go. Mobility implies portability and thereby the users and items can initiate a "real-time" contact with e.g. commercial, human beings, devices and other systems wherever they happen to be on the globe. Broad Reach means that e.g. people, animals, buildings and devices can be reached any time and instantly. These two characteristics break barriers of geography and time. They create value-added attributes which will drive the development of mobile and wireless technology and business models to somewhat we just cant imagine today.

Table 3: Attributes, opportunities and benefits
of Mobile and wireless applications

Lindgren 2010 Attributes, opportunities and benefits of mobile and wireless communication Inspired by Efraim Turban 2010

Attributes, and benefits	Opportunities	Examples of benefits
Ubiquity	Refers to the attribute of being available at any location at any given time.	Real-time information and communication independent of user or device location. Easier information access in real-time environment
Convenience	Convenience for users and devices op operate in the wireless environment.	Rapid progress, easier and faster access to information.
Instant connectivity	Possibility for users and devices to connect easily and quickly to the internet, intranets, other mobile and wireless devices, and databases.	Access to information and knowledge
Personalization	Enables information for individual users, devices purposes.	Purpose and object related information
Location of products, services, human beings, animals, devices and business models	Enables knowledge about where anybody and anything at any particular moment.	Targeting of anybody and anything

Different people and business will define the new opportunities and implication of the Mobile and Wireless communication field in dissimilar and differentiated ways. This is one of the major potential of mobile and wireless communication.

In order to simplify things however, it would probably be easier to separate the question we would like to touch upon in this chapter – namely focusing on two main components: firstly by looking into:

• How the concept of business model has been defined until now? – As is business models and later on

* What we can expect the 'business model' definition and the functions of the business model in 2020 will come to look like - To Be – business models.

Towards an Understanding of the Mobile and Wireless Communication 'Business' Model 2010 – From "As is Business Models" to "To Be Business Models"

Derek F. Abell in his well known 1980 book: "Defining the Business", argued that a business could be defined according to a three dimensional framework:

1. Customer Functions: customer functions rendered by the company
2. Customer Groups - customer groups served by the company
3. Customer Technology - customer technology used to produce customer functions and serve customer groups

Abell argued that a company's core business, the Strategic Business Unit (SBU), was related to what is inside the box (figure 1). This is relzated to what the company could defined as its core business. Abell argued that it was particularly important to the company to know about and define the company's core business because of the strong relation of the core business to the company's mission, goals and strategy – the strategic perspective.

The SBU box represented therefore the core business of the company. The space external to the box represented at that time the potential for innovation and further business development. Abell argued that a company, in general, should stick to its core business (core business model). By doing so, the company could achieve optimisation of business processes and resource utilisations (operational effectiveness).

Derek F. Abell defined the SBU related to a physical business model world – because at this time Mobile and Wireless communication technology hardly existed.

Abell also argued that when innovating out of the core business area, Abell advised the focal company to carefully analyse its own competences and resources. Abell argued that moving into a new strategic new business field (the customer

Figure 1: Three Dimensional Framework for the
'Business' Definition (Abell 1980)
Source.: Derek F. Abell 1980

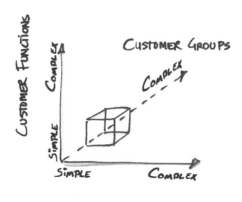

function, customer group or customer technology) would cause a demand for other resources and new competences and, thereby, call for innovation into the field of product development, market development or different degrees of other diversifications.

However, Although Abell's model operated with the three dimensions - which indeed covers and prioritizes some of a company's BM components, it explained very little about how the business [model] structural design really looked like, how it was organized and how it functioned. Why because he was not able to "see" the business model. He was just able to imagine how it could look like and still in an physical context.

Definition of the 'Business Model' Concept
The term 'business model' (BM) became popular in the mid-1990s during the 'dot com era' and up in the 2000s several academics wrote about how the business model looked like and how it could be innovated in a incremental and radical perspective. Some also argued how the business model could be opened (Chesbrough 2008)

Table 4: Examples of incremental and radical building block innovations
Source.: Adapted from Taran, Boer Lindgren 2010

Building block	Incremental innovation 'Do what we do but better'	Radical innovation 'Do something different'
1. Value proposition	Offering 'more of the same'	Offering something different
2. Target customer	Existing market	New market
3. Customer relations	Continuous improvement of existing relationships	New relationship platforms (e.g. physical, virtual, personal)
4. Value chain architecture	Exploitation (e.g. internal, lean, continuous improvement)	Exploration (e.g. open, flexible, diversified)
5. Core competencies	Familiar competencies (e.g. improvement of existing technology)	Disruptively new, unfamiliar competencies (e.g. new emerging technology)
6. Partner network	Familiar (fixed) network	New (dynamic) network (e.g. alliance, joint-venture)
7. Profit formula	Existing processes generating revenue followed by, or accompanied with, incremental processes of retrenchment and cost cutting	New processes to generate revenues followed by, or accompanied with, disruptive processes of retrenchment and cost cutting

Due to today's 'hypercompetition' (D'Aveni 1994) in a globalizing world, companies in all industries worldwide find themselves competing in ever changing environments. Those changes force companies in 2010 to rethink their operational business models more frequently and more fundamentally, as innovation based solely on single business models, just new products and aimed towards just the local markets is no longer sufficient to sustain their competitiveness and survival. Global competitors can relatively easily copy single business models, products and local market segments today and quickly capture business located elsewhere on the globe.

Figure 2: A three-dimensional BM innovativeness scale
Source.: Adapted from Taran, Boer Lindgren 2010

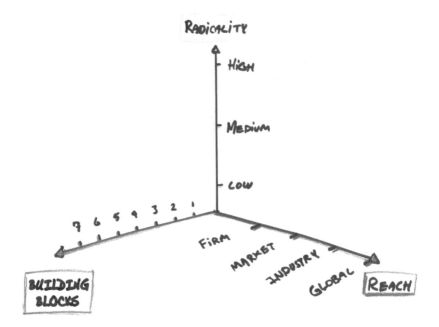

As mobile and wireless communication technology and business ecosystems emerged many companies have started in 2010 to rethink their business model and business structure by shifting to include mobile and wireless communication into their business model on an 'E-form' business basis (Moore 1998). Companies saw during the 2000's the physical and E-form business model as separated – and many in 2010 still do. However more companies will turned out to see the two business models as integrated and it seems that most companies agree that the business model through 2010's will simply be a combination of the two: a physical and a digital 'business model'.

Accordingly, a company's 'business model' in 2010 serves as a building platform that represents the company's operational and physical manifestation but in 2020 it will serve as a building platform that represents the company's physical and digital operational manifest.

Thus, the challenge for the 2010's business model "designers" is firstly to identify the key elements and the key relationships between the physical, the digital and the virtual business model and secondly to describe and form the company's 'To - Be' business model as an integration of the 3 - before innovating it.

Components of 2020 Business Models

As part of the ongoing process of defining the physical, digital and virtual business model concept up to 2020 integrated with mobile and wireless communication technologies, it seems obvious that many companies will try to attempt and clarify the components and building blocks of such as well as the construction of a generic BM with an old perspective and view of business models. They will start thinking the business model with old business models framework – with defining it form a physical and single business model perspective. However because of the mobile and wireless communication technology development they will soon jump to the conclusion - that business models have to be thought about in other ways – with a multitude of dimensions – with a multi business model perspective.

Figure 3: The multitude dimensions of 2020 business models
Source.: Lindgren 2010 inspired by taran, boer and lindgren 2010

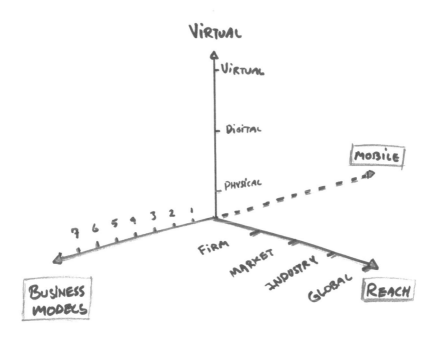

Companies business in 2020 will consists of more business models and more layers of business models. They will differentiate to previous seen business models because they will have and consist of several layers of business models – the multi business model approach. A rather simple example of this will be a mobile company doing one business model on one customer and on the same data another business model to another customer – two revenue (R) streams but with just one cost (C) structure – R1 + R2 – C1 = Profit Formula.

The business models will also consist of both a physical, digital and a virtual layer (Lindgren 2010), which means that e.g. the product or service will have a physical content (clothing and sales service), a digital content (the digital data on the clothing and extra service on e.g. the webside) and finally the product will just exist if there is a need for it – the virtual layer. All these 3 layers will be possible business model layers.

By definition all business models will in 2020 bee reachable and globally. Everybody, wherever and whenever will be able to reach the business model either physical, digital or virtually. The global market will be tremendously big for most business models - the long tail (Andersson 2008) theory will come into real life.

However another - and quiet interesting and new to many companies – business model dimension will be that business models in 2020 will always "be on the run" – they will be mobile.

The product lifecycle and customer lifecycle will as such be an integrated part of the business model. Products and services will be covered with sensors and sensor techniques. This means that products and services can be "tracked and traced" where ever, whenever and by who ever. A new business model potential which adds new possibilities to business models – what we called "the real time process dimension".

An overall view on 2020 business models will be that the BM will by definitions be interrelated business model frameworks with an integrated dimension of mobility. This will indeed set new standards to our understanding of business models. Companies will try to build their business models from what they know about in 2010 - what they called 'a unified perspective of business models' but they will soon come to the conclusion that this is not a possible road to drive and they have to change their mindset about business models and business model innovation when they take the real time process dimension in to consideration.

The indication for 2020 business models is that they will also have a cross-model perspective and there will be no single theory that can fully explain the value creation potential of these multi business models.

Consequently, it is very complex to identify a holistic business model and the building blocks of the business model framework for a mobile and wireless communication business model. However, we believe that they will be as an interactive framework between physical, digital and virtual layers of building blocks with multitude of business models placed simultaneously on many firms at a global level and with the mobile and wireless component built in.

Corresponding with the previous findings of business models - Morris et al., Osterwalder, Pigneur and Tucci (2004), Morris, Schindelhutte and Allen (2003), (Stähler 2001; Weill and Vitale 2001; Petrovic et al. 2001; Gordijn 2002; Afuah and Tucci 2003; Tapscott et al. 2000; Linder and Cantrell 2000; Hamel 2000; Mahadevan 2000; Chesbrough and Rosenbloom 2000; Magretta 2002; Amit and Zott 2001; Applegate and Collura 2001; Maitland and Van de Kar 2002), Christensen (2008); Lindgren, Taran and Boer 2010 nobody have yet come near to 2020 multi business model framework.

Is the Mobile and Wireless Communication Business Model New?

Having a conceptualization of what the Mobile and wireless communication business models can be about - another thing is how to innovate them. According to previous research Magretta (2002), new business models have until now been variations on a generic value chain underlying all businesses, which eventually can be divided into two categories:

1. All the activities associated with production; e.g. designing, purchasing and manufacturing.
2. All the activities associated with selling something; e.g. finding and reaching customers, sales transactions and distributing the products/services.

For that reason, according to her, a new business model can be seen as a new product for unmet needs (new customer segment), or it may focus on a process innovation and a better way of making/selling/distributing an already proven (existing) product or service (to existing and/or new customer segments). Or, formulated more generally, a business model until now have been define as new if one of the "building blocks" is new.

According to Amit and Zott (2001), business model innovation refers not only to products, production processes, distribution channels, and markets, but also to exchange mechanisms and transaction architectures. Therefore they proposed to complement the value chain perspective by concentrating also on processes that enable transactions. In view of that, they concluded that business model innovation does not merely follow the flow of a product from creation to sale, but also includes the steps that are performed in order to complete transactions. Therefore, the business model as a unit of analysis for innovation potentially has a wider scope than the firm boundaries, since it may encompass the capabilities of multiple firms in multiple industries. Also Chesbrough (2007) and the IBM global CEO Study (2006) emphasized the importance of business model innovation to appear in the form of organization structure and network relationship changes, such as alliances, joint-ventures, outsourcing, licensing, and spin-offs.

Accordingly, we (Lindgren, Taran and Boer 2009) concluded that any change can rightfully be called a business model innovation, but some changes are more radical and/or complex than others, and some (e.g. radical product innovation, incremental process improvement) reach farther and are less well understood than others (e.g. a holistic, new to the world departure from all business models known so far). Consequently, we get around the eternal discussion of when a BM innovation is indeed radical or incremental, simple or complex, far reaching or not, and, rather, portray the space in which any business model innovation can be positioned in terms of its innovativeness, defined as a composite of radicality, complexity and reach.

However the mobile and wireless communication business model concept was not in this context taken into consideration. Referring to figure 3: A four-dimensional BM innovativeness scale - the mobile dimension – and some would also call it - the process dimension – has now been taken into consideration.

Levels of Mobile and Wireless Communication Business Model Change

The eternal question in the business model innovation literature is: when can we call something a business model innovation? Is it a new business model if it is new to the company but not within the industry as such? Is it new if the business model is used elsewhere but not in the local industry? Or should it be completely new to all, like e.g. the mobile and wireless communication shopping?

The debate on defining incremental or radical business model innovation (Taran, Boer and Lindgren 2010) concerned the 'how new' question. Incremental business model innovation involves making small-steps on the number of building blocks change - the complexity, the radicallity dimension – defined as the dimension of change of the different building blocks (low, medium, high) and reach (firm, market, industry and world) parameter.

This framework model mainly used existing knowledge to improve, for example, existing products/services, customer practices and structures. It was very short term based and it focused on the focal company performing better than it already did. Radical business model innovation, also called quantum leap innovation, is however the ability to develop new knowledge and may even be based on competences of many companies – the network based business model (Lindgren, Taran and Boer 2009). It does not possess at the start of the process and result in products/services that cannibalize or remove the basis of existing products/ services (Tidd et al. 2005), but address new dimensions of all building blocks at both a physical, digital and virtual layer.

Skarzynski and Gibson (2008) argued that in order to understand how to innovate a business model, you first need to unpack it into individual components and understand how all the pieces fit together in holistic way. Furthermore, according to them, in order to build a breakthrough business model that rivals will find difficult to imitate, companies will need to integrate a whole series of complementary, value creating components so that the effect will be cumulative. Well this is exactly what the mobile and wireless communication business model innovation is about in 2020. This will take the business model innovation, suggested by Linder and Cantrell (2000) right to the level of radicality of business model innovation and presents what they called change models characterized as the extension models and journey models.

Extension models include radical changes by developing new markets – integrated physical, digital and virtual markets, new value chain and value chain functions with again physical, digital and virtual layers, and product/service lines which is both physical, digital and virtual. Most interesting considering the last building block in our 7 building block business model (Taran, Boer and Lindgren 2010) – the profit formular – will both include physical digital and virtual money. The journey model has taken its begin - involving a complete transformation of original business models - where numerous of companies moves deliberately and purposefully to new operating multi business models. The enabler to this development is indeed the mobile and wireless communication technology, which is believed in 2020 to enable and support this evolution.

Open Mobile and Wireless Communication Business Model

Chesbrough (2007) introduced a whole new way of thinking about business models and innovation of business models in his book Open Business models – How to Thrive in the New Innovation Landscape. He argued towards open innovation (2005) and then to opening companies' business models (2008). Until then, all academic work (Abell 1983, Linder and Cantell 2000, Morris 2003, Magretta 2000, Osterwalder et all 2004) and practically all business thinking had been related to a 'closed' business model – sticking to your core business – where the business model was strongly related to a single and focal company. Chesbrough argued that 'a closed business model' was not efficient for future innovation and a global competitive environment. Much innovation was never even related to the core business model and companies who could have used the results of the innovation were prevented from this because of patents, unwillingness to network and even by lack of knowledge by those who had innovated.

Chesbrough argued that, in the future, companies would have to open up their business models and allow other companies to integrate them with their own business models and even take parts out of the model to use it in other business models. These innovations could be extremely valuable and more effectively used in other business models by other companies.

Because of 'a new global business environment' integrated with mobile and wireless communication technologies we claim that Chesbrough's theorem in 2020 will be taken even further. The business models of companies, their products and services, customers, customer relationships, key resources, competences, value chain and profit formulas (Taran, Boer and Lindgren 2010) will have a much deeper structure than Chesbrough ever though about back in 2008. The business models to come will be managed in a much more open way than Chesbrough proposed and they will in 2020 have more and much more open layers than Chesbrough thought about. Developed and registered patents will be open to other companies or other business models of interest in order to support and develop other business models or the innovation of new multi business models.

Previously developed innovations, IPR and competences will become much more value adding building blocks to both existing and new business models in 2020 because of the mobile and wireless communication technology.

Companies will play by different rules in 2020 than in 2010 using different and many business models simultaneously. This open approach – "internet in the clouds" – with "internet in all things" and later "human beings everywhere" will give new opportunities and challenges for both 'giver' and 'taker' of business models and business model innovation. We believe tt will also completely restructure the theory of mobile and wireless communication and related business models leading to a network oriented business model level. Networks - that at the same time, are physical, digital and virtual. These new multi business models will have more different partners' value equation included and will be completely open to all network partners of interest, both within and outside the value chain.

The open business innovation model has thereby in 2020 become not just a true and important concept for further work and development of business models – but has in major parts also be realized.

Mobile and Wireless Communication as a Key Enabler to New Business Models

Mobile and wireless communication technology we know already today will be of ever increasing importance to innovation of new business models and as the backbone of the new business models in 2020. The business models in 2020 will be based on mobile and wireless technology and operates within a mobile and wireless based market environment. This means that pretty much everything will be able to communicate with everybody due to advanced sensor technology and advanced wireless communication technology. Hereby the body of a human being, animals, devices, buildings, clothings e.g. will be able to communicate each other.

This opens for new business models. In 2010 we have only very limit possibilities to communicate from inside a body to outside a body of a human being. Pace makers and bigger instruments put attached to the body is today some of our possibilities but in future 2020 we will be able to communicate to and with practical every part of the body both on small and large scale perspective – organs, legs, arms, blood, brain can be objects for communication and measurements.

Digital devices of our "brains" "intellectual capital" – or devices that have learned our habits, interest and network partners will on an intelligent way communicate for us and thereby help us in a user centric world.

Virtual "avatars" of us and our business models including more advanced secondlife. com, google.com, facebook.com, linkedin.com and World of Warcraft liked environment will "play" or be allowed to "play" our virtual person and virtual life.

In a short term perspective, these business models will run at cost, speed, energy consumption vice, performance and change in the global market that are far above what we have achieved and would believe is possible today.

The development will reach a point of no return in 2020, where a even strong focus on mobility and "wireless life" of human beings and devices to cope with the increasing demand of slipping out of time pressure, being more agile, more flexible is the trend. Business models which are independent of time, place and

Figure 4: 1st generation of physical, digital and virtual Business model based on networks.
Lindgren 2010 inspired by Whinston, A. B. Stahl, D.O., and Choi, 1997

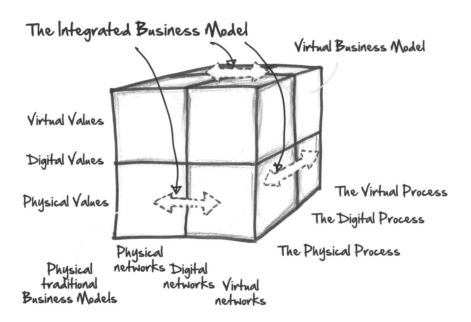

people will be at the 1th. Generation. Integrated business models between physical, digital and virtual business models will be realized .

However we expect more advanced business models to come but it will probably not be standardized in 2020 we believe.

Mobile and Wireless Communication a Move toward a Process Enable New Business Models

An interesting synergy and spinoff of the above mentioned model is a first step towards a a completely new and change in our understanding of the business model – taking us from a static understanding and measurement of the business model to a process based business model.

Business models in 2020 will not only consist of physical, digital and virtual layers but are also based on a process – integrated, connected, mobile, wireless in a continuously process of change - where ever and whenever the user or the customer demands it. All of "the 7 building blocks" in the business model will be in process. We predict that all building blocks can and will change simultaneously in the processes – both individual and across and together – creating continuously interesting new multi business models.

This is both a real challenge of the future multi business models but also the real potentials of multi business models above 2020.

In a long term perspective continuous improvement, learning and innovation (Nonaka & Tacheki 1996 and Bohn & Lindgren 2003), in combination with the continuous need to develop new multi business models, increases the importance of standards, security, legal rights and ethics of multi business models. The ability to include and create mega information fast, flexible and dynamically is not a question of whether it is possible. This will be the innovation platform to every company and network.

So – the answer to the question - What will be the global business models in 2020 – is that practical all business models of the future is global. We expected that all future business models will mainly be based on mobile and wireless communication technologies and will thereby be reachable from all parts of the globe. The majority of these future business models will include both physical, digital and virtual business components.

Discussion

By glancing through the overall research and practice that have been done so far in business model related to the mobile and wireless communication phenomena shows a rather fuzziness of the 'business model' concept and a very simple and single business model perspective. No matter why the conclusion is very simple – we are standing in front of a revolution of our understanding of the business model. Mobile and wireless communication technology have created several new layers and several new dimensions to our previous understanding of the business model.

Chesbrough commenced the trend on open business model but the dimensions of the mobile and wireless communication enabled business models to our opinion cannot be included in his framework. Our current research on business models and innovation has to be reinvented.

From a theory development perspective it is at the previous moment – 2010 not possible to agree on one argument what the phenomena will look like in 2020. Accordingly, in order to provide a solid ground for the multi business model [theory building] research to be built up, and based on the similarities/differences between the various explanations and perspectives of the multi business model, there is a need to narrow down and sharpen the large variation of potentials of mobile and wireless communication, as well as to develop new concepts, ideas and categories to the term 'multi business model' in general, and its innovation processes in particular.

Firstly, as to the components of, what we propose it to be in 2020, the core of the business model, despite the large variation in opinions we can still identify a strong integration between the different components. We proposed a multitude dimensional model including radicallity, reach, complexity and mobility as core dimensions for characterizing the business model to come in 2020.

The next issue concerns the question: when and what can we expect of the new models of business models in 2020? Three approaches have been proposed in this chapter. The first approach 'defines' the 2020 business model as a multi business model with integrated physical, digital and virtual parts. A multi business model in continuously change in any of the building blocks or the relationships between them as a form of dynamic multi business model innovation could be considered. The form and number of building blocks are changed simultane-ously which we consider as the most radical form of business model innovation we expect will become in 2020.

If we combine the 'scale' of radicality (new to an individual, group, company, industry, world), and mobile a four -dimensional space emerged, which does not perhaps provide a precise definition of the multi business model we can expect in 2020, but helps indicate the radicality or, perhaps rather, the real complexity of business model and business model innovation in 2020.

This more accurate language for discussing mobile and wireless communica-tion business model or better enabled business models will help us to better understanding of the the managerial challenges and issues involved using existing innovation and change theories. Not much of previously understood business model theory will be valuable I understanding business models in 2020. Maybe just the building blocks can be transferred – but even our previ-ous customer definition and definition of money – and the profit formula - has to be revised.

Building a Process to Mobile and Wireless Communication Multi Business Model Innovation

Our findings so far has illustrated that the multi business model phenomenon is not well understood until now and innovation of these enabled by mobile and wireless communication technology is just partly understood, and the combina-tion of the two leads to a high level of uncertainty and fuzziness, particularly to many company managers who will consequently be faced with tremendous challenges and obstacles when attempting to successfully innovate their multi business model (i.e. financial obstacles, strategic [choices] obstacles, operational obstacles, cultural obstacles, etc.).

Given this, and the understanding of radicallity in multi business model innovation seems to be indeed a 'risky business', is leading to the following tentative theory statement, namely: Implementation of risk management processes within the overall innovation process might reduce the level of uncertainties, overall costs and [implementation] time, when developing busi-ness models of 2020.

Futuristic Outlook on Emerging Global Mobile and Wireless Communication Business Models In 2020

The business model concept will continue to evolve and embrace new perceptions, challenges and opportunities. As part of the authors' preliminary research, the following important trends and characteristics were found related to emerging globale mobile and wireless communication business models in 2020.

Table 5: Future trends to Mobile and wireless BM's Innovation

Context for Innovation	2010	2020
Market	Commencing global markets Unstable Mainly physical but commencing digital and virtual market	Fully global market Fragmented, Dynamic, Customised, mobile and process oriented New markets (Blue Ocean) Integrated physical, digital and virtual markets
Technology	Single technology Expensive Data power low Stable	Mix of integrated mobile and communication technology or multi-technology based Satellite based Sensor based 4G and preliminary 5G based Data power capacity challenged Stable technology Standardised technology Security technology based Stable new technology changes
Network	Closed networks, local networks, fixed networks	Open physical, digital and virtual networks Dynamic networks, Integrated physical, digital and Virtual networks, Global physical, digital and virtual networks

Context for Innovation	2010	2020
Companies' Competencies	Unstable competencies under development inside primary companies or in narrow networks	Dynamic – flexible and virtual competencies Competencies continuously under development Competencies developed within a multitude of network and together with network partners - shared core competencies and skills in the innovation process
Products	Mostly physical products to some extent immaterial products Unstable product – Short life cycle Limited distribution and marketing channels	A mixture of physical, immaterial, digital and virtual products and services – however with an overweight to digital and virtual products Continuous development of physical, digital and virtual products and service - short life cycles of products and service Multitude of physical, digital and virtual distribution and marketing channels
Business model Innovation process	Unstable models Slow, linear business model innovation process	Many business model innovation models (flexible models, dynamic models, learning by doing, using, interacting). Rapid prototyping models of business models Lean business model innovation process
Success Criteria	Individual success, business model innovation speed, business model time to market, business model cost and performance, business models to the local market Emphasis on short term success criteria. More emphasis on continuous improvements and managing tangible assets efficiently.	Network-based business model success, right-speed business model innovation, business model time to market, business model cost and performance optimization, global business model markets Emphasis on sustainability in business models- short and long term success criteria are integrated. More emphasis on radical business model innovation Most emphasis on managing intangible assets efficiently in business models Focus on Innovation leadership

Table 5 above is clearly demonstrating that the changes in the economy will be the rules of the game and companies must in the future learn to adjust their mindset to a multi business model perspective.

The new multi business model based economy is fast coming. It is highly user and customer driven. Globalisation are being strongly related to network and differentiation, learning and continuous innovation considered to be implemented via continuously new and open network formations.

Mobile and Wireless Communication Business Model Innovation Space in 2020

The dimensions of multi business model innovation space, presented earlier can be further analyzed into possible expected outcomes, when companies are considering how to innovate their multi business models, which eventually can result in [roughly] eight possible multi business model innovation outcomes.

Table 6: Mobile And Wireless Communication Business Model Innovation Space In 2020

Radicality	New to whom	Number of building blocks changed	Expected business model outcome
High	Low	Low	**'Me too' BM.** Follow the others. Second mover strategy. Radical new value, but only to the organization.
High	Low	High	**'Me too' BM.** Despite the radical internal change processes, it is still a second mover strategy.
High	High	Low	Potentially, **Highly competitive BM.** Involved relatively low change management (and risks) processes to be implemented.

Radicality	New to whom	Number of building blocks changed	Expected business model outcome
High	High	High	**Radical new BM.** Very risky, but also very competitive (blue -ocean)
Low	Low	Low	**Laggard BM.** Not perusing any radical innovation, focus is on continuous improvements, risk of staying behind competition for the long run
Low	High	High	Potentially, **Highly competitive BM.** Requires high levels of change management (and risks) processes to be implemented.
Low	Low	High	**Laggard BM.** Not perusing any radical innovation, stay behind competition for the long run, and also highly risky due to high levels of incremental innovations and change management processes that are needed to be implemented simultaneously
Low	High	Low	Potentially, **Highly competitive BM.** Involved relatively low change management (and risks) processes to be implemented.

Future Thinking is Also About Network-Based Ict Enabled Business Model Platforms

Competition in 2020 will not be single company based and single business model will hardly exist. Consequently various ways of open network based multi business model created in network-based innovation processes will be the standard. A multitude of external competences and knowledge zones (Amidon 2008), among others, will be involved together with magnitude of customers and suppliers, knowledge consultants, venture capitalists and competing networks. Open network based innovation assist to better leverage internal and external capabilities in the networks, as well as in opening continuously new markets and continuously experimenting with new cutting edge technologies and markets. Consequently, many networks are finding themselves increasingly tied to other networks either physical, digital and or virtual or all 3 integrated. The challenge each of them faces is to adjust their multi business model continuously meet their networkpartners' core competences and multi business models' value, so as to create a new platform for collaboration and innovation. This tendency, if continued, will have tremendous consequences on the way that we will see business, the company terminology – if it will keep on existing in 2020. Competition in 2020 will take new dimensions and we will learn new dimensions of competition.

Consequently, open multi business model innovation initiatives are expected to grow significantly from now to 2020 in volume (i.e. joint ventures, spin-offs, licensing, outsourcing, and alliances). Open multi business model innovation, which can also be considered as network based business model innovation platforms, may therefore in 2020 either result from most possible equal partnership, or take place within loose, and more diversified physical, digital and virtual networks. In either case, some network partners will be forced to change their individual business model logic in the time to come - some more radically than others. In addition, the development of a network-based business model innovation may result in a radical change of our understanding of the customer, the customer and product lifecycle, since they involve and are integrates in the mobile and wireless technology, with new and a magnitude value propositions, new and magnitude of value chains, new and magnitude of network formations, and new markets.

Furthermore, the new network based market formation ties will not be characterised by industrial homogeneity, but rather by the large diversification of its network partners' identity for the purpose of pursuing radical multi business model innovation possibilities.

All in all, the present research and description has led the researchers to believe that the business model concept until 2020 will continuously evolve and embrace new perceptions, challenges and opportunities – beyond our todays imagination.

Conclusion

Today ICT based communication and technology is getting even more popular and important everyday. In this chapter we briefly introduced an overview of the evolution and the background of 1g to 5g, compared the differences and illustrated how 5g may work for business models in the 2020. 4g just right started and there is still many standards and security technologies to come, which are still in developing process. Therefore, no one can really be sure what the future 5g will look like and what services it will offer to people. However, we can get the general idea about 5g from academic research; 5g is the evolution based on 3g's and 4 g's limitation and it will fulfill the idea of wwww, world wide wireless web, offering more services and smooth global roaming with low cost.

A lot of buzzwords have come and gone in time, but it seems the business model concept will change tremendously until 2020. Despite its fuzzy definition and operationalization, it is capturing more and more attention of academics as well as company managers.
The objective of this chapter was to build a better understanding to global ICT based business models in 2020.

Through the chapter, we introduced and illustrated a framework of future and emerging new business models to come in 2020 and together the implementation process together with the challenges of these BM innovations.

It is the researchers' belief that the future BM concept, particularly with relation to its innovation will be one of a multi business model. A new understanding and theory of business models and what they are all about seems to be developing quickly at present. It is the researchers believe that the future BM concept should be considered and analyzed from a network perspective. A more network based business model moving from a closed to an open business models is developing, with much more focus on values, processes, cost - not only of partners inside the multi business models but also those outside – network partners who are either in the first, second or even third levels of the chain of the multi business model. Multi business models which are integrated physical, digital, and virtual models.

So when addressing a futuristic outlook on emerging multi business models and emerging business models based on mobile and wireless communication technology - we need more research about the multi business models and how these will look like. This concerns both a theoretical and a practical perspective. Some short cases will point to some future examples and possibilities of multi business models.

The aim of this chapter was to open the discussion on multi business model and multi business model innovation as a important field of research, as well as in drawing future potential innovation possibilities, and expected outcomes, when considering how to innovate companies' business models.

References

- Abell, D.F. (1980), *Defining the Business: The Starting Point of Strategic Planning,* Prentice-Hall.

- Afuah, A. and Tucci, C. (2003), *Internet Business Models and Strategies,* Boston: McGraw Hill.

- Amit, R. and Zott, C. (2001) *Value creation in e-business,* Strategic Management Journal, Vol. 22, Nos. 6–7, pp. 493–520.

- Anderson Chris (2008) *The Long Tail Hyperion,* ISBN 978 – 1 – 4013-0966-4

- Applegate, L.M. (2001) *E-business models: Making sense of the internet business landscape,* in G. Dickson, W. Gary and G. DeSanctis (Eds.), Information Technology and the Future Enterprise: New Models for Managers, Upper Saddle River: N.J. Prentice Hall.

- Baker, M. and Hart, S. (2007) *Product Strategy and Management,* Harlow: Prentice Hall, pp. 157–196.

- Barney, J. B. (1991) *Firm resources and sustained competitive advantage.* Journal of Management, 17: 99–120.

- Burt, R. S. (1992) *Structural Holes: The Social Structure of Competition.* Harvard University Press: Cambridge, MA.

- Carlile, P.R. & C. M. Christensen (2005) *The cycles of theory building in management research Version 6.0.* Working paper, Innosight, www.innosight.com/documents/Theory%20Building.pdf,

- Chesbrough, H. and Rosenbloom, R.S. (2000) *The role of the business model in capturing value from innovation: Evidence from XEROX Corporation's technology spinoff companies,* Industrial and Corporate Change, Vol. 11, No. 3, pp. 529–555.

- Chesbrough, H. (2006) *Open Business Models. How to Thrive in the New Innovation Landscape,* Harvard Business School.

- Christensen, C. M. (2006) *The Ongoing Process of Building a Theory of Disruption.* Journal of Product Innovation Management 23: 39–55.

- Cooper, R. (1993) *Winning at New Products: Accelerating the Process from Idea to Launch,* 2ed edition, Boston, Addison-Wesley.

- Doz, Y. and Hamel, G. (1998) *Alliance Advantage.* Harvard Business Press: Boston, MA.

- Dyer, J. and Singh, H. (1998) *The relational view: cooperative strategy and sources of interorganizational competitive advantage.* Academy of Management Review, 23: 660–679.

- Gordijn, J. (2002) *Value-based Requirements Engineering - Exploring Innovative E-Commerce Ideas*, PhD thesis, Vrije Universiteit Amsterdam, The Netherlands.

- Green Tech Media (2010).

- Hamel, G. (2000) *Leading the Revolution,* Boston: Harvard Business School Press.

- IBM (2006) *Expanding the Innovation Horizon*, http://www-935.ibm.com/services/uk/bcs/html/t_ceo.html

- IBM (2008) *The enterprise of the future,* http://www-935.ibm.com/services/us/gbs/bus/html/ceostudy2008. html

- Kuhn, T. (1962) *The Structure of Scientific Revolutions.* Chicago: University of Chicago Press.

- Leifer, R. (2002) *Critical Factors Predicting Radial Innovation Success,* Berlin: Technische Universität.

- Linder, J. and Cantrell, S. (2000) *Changing Business Models: Surveying the Landscape,* Cambridge: Accenture Institute for Strategic Change.

- Lindgren, P., Taran, Y. and Schmidt, A.M. (2008) *The analytical model for NEWGIBM,* in P. Lindgren and E.S. Rasmussen (Eds.) New Global ICT-Based Business Models, Aalborg Universitetsforlag.

- Lindgren, P., Taran, Y. Boer., H. (2009) *From Single Firm to Network-Based Business Model Innovation.* the 'International Journal of Entrepreneurship and Innovation Management' (IJEIM)

- Loukis, E. and Tavlaki, E. (2005) *Business Model: A Perquisite for Success in the Network Economy, BLED Proceedings.*

- Magretta, J. (2002) *Why business models matter?,* Harvard Business Review, Vol. 80, No. 5, pp. 86–92.

- Mahadevan, B. (2000) *Business models for internet-based e-commerce. An anatomy,* California Management Review, Vol. 42, No. 4, pp. 55–69.

- Moore, J.F. (1998) *The new corporate form,* in D. Tapscott, A. Lowy, and D. Ticoll (Eds.) Blueprint To The Digital Economy – Creating Wealth in the Era of E-Business, New York: McGraw-Hill.

- Morris, M., Schmindehutte, M. and Allen, J. (2003) *The entrepreneur's business model: Toward a unified perspective,* Journal of Business Research, Vol. 58, No. 6, pp. 726–735.

- Osterwalder, A., Pigneur, Y. and Tucci, L.C. (2004) *Clarifying business models: Origins, present, and future of the concept,* Communications of AIS, No. 16, pp. 1–25.

- Paavilainnen (2002).

- Petrovic, O., Kittl, C. and Teksten, R.D. (2001) *Developing business models for e-business,* Proceedings of the International Conference on Electronic Commerce, Vienna, October-November.

- Porter, M.E. (1985) *Competitive Advantage: Creating and Sustaining Superior Performance.* Free Press, New York.

- Porter, M.E. (2001). *Strategy and the Internet,* Harvard Business Review, (79)3 pp. 63–78

- Printed Media (2010)

- Rogers, E. M. (1983), *Diffusion of Innovations,* 3rd edn, The Free Press, New York.

- Rosenau, J.N. (1993). *Turbulent change,* in P. R. Viotti and M. V. Kauppi (Eds.), International Relations Theory: Realism, Pluralism, Globalism, New York: Macmillan Publishing Company, pp.438–448.

- Schumpeter A.. J. (1942), *Capitalism, Socialism and Democracy.* New York: Harper & Row.

- Skarzynski P. and Gibson R. (2008) *Innovation to the Core,* Boston: Harvard Business School Publishing.

- Stahler, P. (2001) *Geschäftsmodelle in der digitalen Ökonomie. Merkmale, Strategien und Auswirkungen,* PhD thesis, University of St. Gallen, Switzerland.

- Stewart, D.W. and Zhao, Q. (2000) *Internet marketing, business models, and public policy,* Journal of Public Policy & Marketing, Vol. 19, pp. 287–296.

- Taran, Boer Lindgren 2010 *Managing risks in business model innovation processes,* CInet Zürich 2010.

- Tapscott, D., Ticoll, D. and Lowy, A, (2000) *Digital Capital. Harnessing the Power of Business Webs,* Boston: Harvard Business School Press.

- Tidd, J., Bessant, J. and Pavitt, K. (2005) *Managing Innovation. Integrating Technological, Market and Organizational Change,* Chicester: John Wiley & Sons.

- Times (2010).

- Timmers, P. (1998) *Business models for electronic markets,* Journal on Electronic Markets, Vol. 8, No. 2, pp. 3–8.

- Turban Efraim 2004, 2006, 2008, 2010, E*lectronic Commerce A Managerial Perspective,* ISBN-13 978-0-13-224331-5

- Ulrich, K.T and Eppinger, S.D. (2000) *Product Design and Development,* New York: McGraw-Hill.

- Weill, P. and Vitale, M.R. (2001) *Place to Space,* Boston: Harvard Business School Press.

- Wernerfelt, B. (1984) *A resource-based view of the firm,* Strategic Management Journal, 5: 171–180.

- Williamson O.E. (1981) *The Economics of Organization: the Transaction Cost Approach,* Am J Sociol 87(4), pp. 77–548.

- Wind, Y. (1973) *A new procedure for concept evaluation,* Journal of Marketing, Vol. 37, October, pp. 2–11.

- Whinston, A. B. Stahl, D.O., and Choi S., 1997 *The Economics of Electronic Commerce,* Indianapolis, IN: Macmillan Technical Publishing

Research Team Presentation
– Positions At The Project Period

Senior Researchers

Associate Professor Peter Lindgren

Associate Professor Erik S. Rasmussen

Professor Anders Drejer

Assistant Professor Niels N. Grünbaum

Associate Professor Svend Hollensen

Assistant Professor Anders McIlquham Schmidt

Associate Professor Per Servais

Junior Researchers and Assistants

Doctoral Research Student Carsten Bergenholtz

Doctoral Research Student René C. Goduscheit

Doctoral Research Student Jacob H. Jørgensen

Doctoral Research Student Kristin F. Saghaug

Doctoral Research Student Yariv Taran

Doctoral Research Student Andreas Slavensky

Research Assistant Kean Sørensen

Peter Lindgren

Associate Professor, Ph.D.
Center for Industrial Production
Aalborg University
Fibigerstræde 16, DK - 9220 Aalborg
Phone: +45 9635 8987
Fax: +45 9815 3030
E-mail: pel@production.aau.dk

Current Positions
Associate Professor Ph.d. in High Speed Innovation and New Business Development at Center for Industrial Production, Aalborg University. Part time lecturer in E-business at University of Aarhus.

Head of the International Center for Innovation[1] (ICI) (2008 - 2012) at Aalborg University. Project Manager for the M-commerce project[2] 2007- 2009, Ministry of Science, Technology and Innovation. Project Manager for the NEWGIBM research project[3] (2005 – 2007), Ministry of Science, Technology and Innovation. Project Manager for the Danish part of the EU Proact project[4] 'EU DG6'. Leader of the 'Global Innnovation' project (2006 - 2007)[5] and Global E-business project[6] (2006 – 2008). Member of the Science and Enterprise Network.

Member of the research group 'The Global Production Group'. Member of the research group PUIN 'Product Development in Networks'. Member of the E-PUIN research group[7]. Board Member of the Knowledge Center for E-Business[8]. Member of the Pitnit research group, Ministry of Science, Technology and Innovation. Member of the Tom research group, Ministry of Science, Italy. Member of the research group 'Research on Entrepreneurship in SMEs'.

Has published several books, conference papers, cases for teaching[9].

Has been involved in the development and teaching of courses in global business development for civil engineers.

Is further the organiser of a large number of workshops with the aim to communicate up to date research on high speed innovation, networkbased innovation, e-business and new global business models.

Research Interests
E-business. High Speed Innovation. Innovation Leadership. New Business Models. Innovation of Global Business Models.

---

1 www.ici.aau.dk
2 www.m-commerce.dk
3 www.newgibm.dk
4 www.proactdk.dk.
5 www.globalinnovation.dk
6 www.globalebusiness.dk
7 www.epuin.dk
8 www.ve-b.dk
9 www.vbn.dk
   www.womeninbusiness.dk
   www.cip.aau.dk/Forskning/Projekter/CenSec/

Erik Stavnsager Rasmussen

Associate Professor, Ph.D.
International Business & Entrepreneurship
University of Southern Denmark
Campusvej 55, DK - 5230 Odense M.
Phone: +45 6550 3370
Fax: +45 6615 5129
E-mail: era@sam.sdu.dk

Current Positions
Associate Professor, International Business & Entrepreneurship -

Researcher in company internationalization and B2B marketing.

Ph.D. from University of Southern Denmark about "Born Globals – companies that are born international.

Research Interests
Born Global Companies and Company branding in industrial markets. Analysis of "Super brands" on B2B markets in Denmark. Method development and – understanding of business economies.

Anders Drejer

Professor
The faculty of Social Science
Aalborg University
Fibigerstræde 4
DK - 9220 Aalborg Ø
Phone.: +45 9940 8279
E-mail: ad@business.aau.dk

Current Positions

Full Professor in strategy and business development at Aalborg University from 2010. Anders Drejer holds a Full Professorship in strategy and business development at Aarhus School of Business and holds a Ph.d. in business strategy and competence development from Aalborg Universitet from 1996.

Anders Drejer was head of the Strategy Lab at Aarhus School of Business (20-2010) at Aalborg University. Membr of the NEWGIBM research project (2005–2007), Ministry of Science, Tech-nology and Innovation.

Anders Drejer has carried out several Research management functions: 1996-1999: Member of the management group for "Center for Technology Management", 1997-1999: Member of the project management of the EU project "Enhancing Competitiveness of SMEs via Innovation", 1999-2003: Member of the management group for "Centre for Industrial Production", CIP, 1999-2003: Pro-gramme manager for Intelligent manufacturing and New Product Development research programmes under "Centre for Industrial Production", CIP.

Anders Drejer has further in 2001 been Key-note speaker at "the 10th International Conference on Management of Technology IAMOT 2001", June 19-22, Lausanne, Switzerland, 2001. He was from 2002-2003 Chairman for the organising committee for the Conference, Managing Innovative Manufacturing (MIM) 2003 and from 2003 Leader of the Strategic Management Group at Aarhus School of Business

Anders Drejer has besides many international reserach publications published several books focused on management and practice. Anders Drejer is a often used speeaker at conference, coach and board member in danish companies.

Niels N. Grünbaum

Assistant Professor, Ph.D.
Danish Institute of Border Region Studies
University of Southern Denmark
Alsion 2, DK – 6400 Sønderborg
E-mail: ngr@sam.sdu.dk

Current Positions
Assistant Professor, Ph.D.

Research Interests
Industrial Marketing B2B market, Qualitative management, Research Methods

Svend Hollensen

Associate Professor, Ph.D.
Department of Border Region Studies
University of Southern Denmark
Alsion 2, DK – 6400 Sønderborg
Phone: +45 6550 1218
Fax: +45 6550 1093
E-mail: svend@sam.sdu.dk

Current Position

Ph.D. and Associate Professor in 'International Marketing' at the Department of Border Region Studies, University of Southern Denmark (Sønderborg).

He is responsible for the development of the Business Relationship Management (BRM) Master level programme at University of Southern Denmark (Sønderborg). Svend combines theoretical insight with working experience, for example as an international marketing manager. He is the author of a number of textbooks on marketing in Danish, e.g. a textbook on marketing planning based on cases.

Has also published several textbooks on marketing in English at Pearson/Prentice Hall (UK). One of these titles 'Global Marketing' has been quite successful and has been translated into other languages, e.g. Russian and Chinese. The fourth edition of this textbook was published in May 2007. In 2008 Pearson/Prentice Hall will publish a new textbook by Svend Hollensen 'Essentials of Global Marketing'.

Svend has also worked as a business consultant for several multinational companies, as well as global organisations like the World Bank.

Research Interests

International Marketing. Internationalisation and export. Marketing Planning. Business-to-Business Marketing. Competence development in relation to internationalisation of companies.

Anders McIlquham-Schmidt

Assistant Professor, Ph.D.
Department of Management &
International Business
Aarhus School of Business
University of Aarhus
Haslegaardsvej 10, DK-8210 Århus V.
Phone: +45 8948 6688
Fax: + 45 8615 7629
E-mail: amc@asb.dk

Current Position
Assistant Professor, Ph.D. at the Department of Management, Aarhus School of Business – University of Aarhus.

He currently teaches at the bachelor and masters level in strategic management and innovation.

Research Interests
Strategy and Business Development. Innovation. Corporate Performance. Performance Measurement. Meta-Analysis.

Per Servais

Associate Professor, Ph.D.
Department of Marketing & Management
University of Southern Denmark
Campusvej 55, DK-5230 Odense M.
Phone: +45 6550 3266
Fax: + 45 6615 5129
E-mail: per@sam.sdu.dk

Current Positions
Associate Professor, Department of Marketing, University of Southern Denmark.

- Program responsible Graduate exchange program in European Business (EMBS)
- Profile responsible Graduate program in International Marketing.
- Program director Flexible-MBA (with ASB)
- European Master of Business Study responsible – SDU
- Study Director/head of study board – Business studies, SDU-Odense

Former member of the steering committee of "Knowledge for Growth" (University Institution to establish connection between private industry and the academic society). Former member of the Venture Cup jury. Appointed by The Ministry of Science, Technology and Innovation as external examiner in Business administration (Marketing and Management). Member of BrandBase and CESFO. Member of Erhvervsklub Fyn. Former member of NOREK (Nordplus-network for Business Administration and Economics in the Nordic area).

Ad Hoc Reviewer: Journal of International Marketing, International Business Review, Journal of International Business, Journal of Supply Chain Management.

Research Interests
The formation and growth of International New Ventures. Industrial firms' international purchasing and sourcing activities. Buying behaviour in small industrial firms. E-business and E-procurement in industrial firms. Branding on Industrial markets. Out- and Insourcing activities in industrial firms. Relationships and de-internationalisation in small Firms. International Entrepreneurship.

Carsten Bergenholtz

Doctoral Research Student
Department of Management
CORE
Aarhus School of Business
University of Aarhus
Haslegårdsvej 10, DK - 8210 Aarhus V
Phone: +45 8948 6702
Fax: +45 8948 6125
E-mail: cabe@asb.dk

Current Position

Doctoral Research student at the Department of Management, Aarhus School of Business - University of Aarhus since August 2007.

Joined the NEWGIBM Research Project in 2006 as a research assistant, Center for Industrial Production, Aalborg University. Contributed to paper presented at The European Conference on Research Methodology for Business and Management Studies 2007. Presented a paper at The Proceedings of the 2nd International ISEOR, 2007.

Research Interests

Innovation. Networks. Methodology. Science and ethics.

René Chester Goduscheit

Doctoral Research Student
Center for Industrial Production
Aalborg University
Fibigerstræde 16, DK - 9220 Aalborg
Phone: +45 2635 0699
Fax: +45 9815 3040
E-mail: chester@production.aau.dk

Current Position
Assistant Professor Ph.d. at University of Southen Denmark.
Ph.D. from Center for Industrial Production at Aalborg University November 13th. 2009.

Presented a paper at The European Conference on Research Methodology for Business and Management Studies. Presented a paper at The 7th International Continuous Innovation Network (CINet). Presented a paper at The Proceedings of the 2nd International ISEOR. Presented a paper at The International Economic Modernization and Social Development Conference, nr. 7, Moskva. Presented two working papers at the IMP Journal seminar in Trondheim, Norway, 2007.

Research Interests
Leadership and innovation in networks. Innovation in inter organisational settings.

Jacob Høj Jørgensen

Doctoral Research Student
Center for Industrial Production
Aalborg University
Fibigerstræde 16
9220 Aalborg
Phone: +45 9635 7105
Fax: +45 9815 3040
E-mail: jhj@production.aau.dk

Current Position
Doctoral Research Student, Center for Industrial Production, Aalborg University since January 2006.

Involved in the Global Innovation project. Presented a working paper on action research in inter-organisational networks: impartial studies or the Trojan horse? at the 6th European Conference on Research Methodoly og Business and Management Studies, 2007. Presented a working paper on collaboration with customers in network-based innovation processes – networks and relations in the Fuzzy Front-End at the third IMP Journal Seminar, 2007. Presented a working paper on customer innovation process leadership at the 7th International Continuous Innovation Network, 2007. Presented a working paper on innovation, product development and new business models in networks: How to come from case studies to a valid and operational theory at University Jean Moulin, Lyon 3 and Research Methods Division, Academy of Management, 2007.

Research Interests
Customer driven innovation processes. Customer driven innovation process leadership. Innovation and product development. Innovation in networks.

Kristin Falck Saghaug

Ph.D Fellow
Center for Industrial Production
Aalborg University
Fibigerstræde 16, DK - 9220 Aalborg
Phone: +45 9635 7100
E-mail: kfs@production.aau.dk

Current Position
Research Assistant at the Center for Industrial Production, Aalborg University
Obtained her master's degree in theology from Aarhus University in 2006. Has
been associated to the NEWGIBM research project as research assistant since
August 2007.

Research Interests
Cross disciplinary studies in the fields of human creativity.

Yariv Taran

Doctoral Research Student
Center for Industrial Production
Aalborg University
Fibigerstræde 16, DK - 9220 Aalborg
Phone: +45 9635 9948
Fax: +45 9815 3040
E-mail: yariv@production.aau.dk

Current Position

Doctoral Research Student at the Center for Industrial Production, Aalborg University since November 1st 2007.

Immediately after having obtained his Master's degree in business economy end of 2006, Yariv Taran joined the NEWGIBM research project where he worked as a research assistant.

Research Interest

Yariv Tarans research interests lie within the field of business model innovation. Other research interests are subjects such as intellectual capital management, knowledge management, entrepreneurship and regional systems of innovation (i.e. clusters, triple-helix, SME's).

Andreas Slavinsky

Doctoral Research Student
Center for Industrial Production
Aalborg University
Fibigerstræde 16, DK - 9220 Aalborg
Phone: +45 21551112
E-mail: as@production.aau.dk

Current Position
Doctoral Research student at the Center for Industrial Production, Aalborg University 2010.

Joined the Center for Industrial Production in 2009 as a research assistant with main contribution and work related to Global Network Operations and Business Models.

Co-author and presenter of paper at the European Operations Management Association Conference 2006. Former production engineer at world leading intelligent lightning manufacture with experience within production optimization, project management, sales & marketing and outsourcing.

Research Interest
Business Models, Global Operations Networks, Value Chains, Internationalisation outsourcing/offshoring & Global Manufacturing Strategy

Kean Sørensen

Research Associate
Department of Management
Strategy-Lab
Aarhus School of Business
University of Aarhus
Haslegårdsvej 10, DK - 8210 Aarhus V
Phone: +45 8948 6721
Fax: +45 8948 6125
E-mail: KeanS@asb.dk

Current Position
Research Associate, Department of Management, Strategy-Lab, Aarhus School of Business – University of Aarhus.

Holds a position as Research Associate at Strategy Lab since 2006. Involved in the NEWGIBM case studies. Network broker for EMBA internet portal. Active in a number of projects such as the development of the EMBA alumni network portal, Innovation Cup 2007, Innovation Cup 2006, Business Development Project for the Danish Greenhouse Industry, Case Development for Innovation Rooms.

Research Interests
Strategic management.

Milton Keynes UK
Ingram Content Group UK Ltd.
UKHW050644141024
449569UK00011B/601